SpringerBriefs in Food, Health, and Nutrition

Springer Briefs in Food, Health, and Nutrition present concise summaries of cutting edge research and practical applications across a wide range of topics related to the field of food science.

Editor-in-Chief
Richard W. Hartel
University of Wisconsin—Madison, USA

Associate Editors
J. Peter Clark, *Consultant to the Process Industries, USA*
John W. Finley, *Louisiana State University, USA*
David Rodriguez-Lazaro, *ITACyL, Spain*
David Topping, *CSIRO, Australia*

For further volumes:
http://www.springer.com/series/10203

C. Anandharamakrishnan

Techniques for Nanoencapsulation of Food Ingredients

 Springer

C. Anandharamakrishnan
Food Engineering Department
CSIR-Central Food Tech. Research Inst.
Mysore, India

ISSN 2197-571X   ISSN 2197-5728 (electronic)
ISBN 978-1-4614-9386-0   ISBN 978-1-4614-9387-7 (eBook)
DOI 10.1007/978-1-4614-9387-7
Springer New York Heidelberg Dordrecht London

Library of Congress Control Number: 2013953883

Printed on acid-free paper

Springer is part of Springer Science+Business Media (www.springer.com)

Preface

Nanoencapsulation is one of the most promising technologies for entrapping bioactive compounds so as to protect them from degradation. Currently, various nanoencapsulation techniques are emerging, each with their own merits and demerits. Nanoencapsulation of food ingredients has versatile advantages for targeted, site-specific delivery and efficient absorption through cells. Furthermore, each encapsulation technique has some unique operating factors, which affect the final outcome of the nanoencapsulates, and those factors need to be investigated and optimized. Cutting-edge research is being carried out on the application of nanotechnology in food ingredients for improved health and industrial significance. Herein, the main focus is on the various techniques used for nanoencapsulation of food ingredients, such as emulsification, coacervation, inclusion encapsulation, nanoprecipitation, lipid-based nanocarriers, electrospraying, electrospinning, freeze-drying, and spray drying. The current state of knowledge, limitations of these techniques, characterization of nanoparticles, applications, and recent trends, together with safety and regulatory issues related to nanoencapsulation of food ingredients, are also highlighted in this brief.

Mysore, India
C. Anandharamakrishnan

Acknowledgments

I am extremely grateful to Prof. Ram Rajasekharan, Director, CSIR-Central Food Technological Research Institute, Mysore, India for his valuable guidance, scientific advice, and continuous encouragement.

I would like to express my sincere gratitude to my guide and mentor Prof. Chris Rielly, Professor and Head, Chemical Engineering Department, Loughborough University, UK and Dr. Andy Stapley, Senior Lecturer, Chemical Engineering Department, Loughborough University, UK for their help and support.

I would like to thank all my Ph.D. students and especially Ms. Ezhilarasi, Ms. Anu Bhushani, Ms. Padma Ishwarya and Ms. Divya for their help.

My heartfelt thanks to my parents and sister for their prayers, love, encouragement, and support right from the beginning. This work would not have been possible without my wife Dr. G. Shashikala and my son A. Nishanth; I appreciate their sacrifice, patience, and moral support throughout my research career.

Contents

Chapter 1
Nanoencapsulation of Food Bioactive Compounds

Nanoscience and nanotechnology are new frontiers of this century and becoming highly important due to its wide application in various fields. Food nanotechnology is an emerging technology being potential to generate innovative products and processes in the food industry. Many reviews and research papers have been published on application of nanotechnology in foods. However, only a few reports are focused on nanoencapsulation of food ingredients. Therefore, the main focus here is to discuss the various nanoencapsulation techniques, their advantages, flaws and variations, as well as to appraise the interesting emerging technologies and trends in this field, along with the safety and regulatory issues.

1.1 Nanotechnology

Nanotechnology is an emerging technology that holds the potential to transform various industries in the world. It deals with the production, processing, and application of materials with a size of less than 1,000 nm (Sanguansri and Augustin 2006). The term "nano" refers to a magnitude of 10^{-9} m (Quintanilla-Carvajal et al. 2010). The British Standards Institution has defined nanotechnology as the design, characterization, production, and application of structures, devices, and systems by controlling shape and size at the nanoscale (Bawa et al. 2005). Nanotechnology has emerged as one of the most promising scientific fields of research in the last few decades. Reduction in particle size to the nanoscale range increases the surface-to-volume ratio, which consequently increases reactivity manyfold with changes in mechanical, electrical, and optical properties. These properties offer many unique and novel applications in various fields (Neethirajan and Jayas 2010). Over the last decade, research in nanotechnology has skyrocketed and there are numerous companies specialised in the fabrication of new forms of nanosized matter, with expected applications in medical therapeutics, diagnostics, energy production, molecular

C. Anandharamakrishnan, *Techniques for Nanoencapsulation of Food Ingredients*,
SpringerBriefs in Food, Health, and Nutrition, DOI 10.1007/978-1-4614-9387-7_1,
© C. Anandharamakrishnan 2014

computing, and structural materials (Duncan 2011). The properties of materials differ when going from larger scale to macromolecular scale, which broadens their application. Research focuses on controlling the morphology, composition, and size of nanomaterials, thus enabling the tailoring of unique and desired physical properties of the resulting materials (Kuan et al. 2012; Ezhilarasi et al. 2013).

Nanotechnology can be applied in all phases of the food cycle "from farm to fork." The food industry is facing enormous challenges in developing and implementing systems that can produce high quality, safe foods while being efficient, environmentally acceptable, and sustainable (Manufuture 2006). To answer this complex set of engineering and scientific challenges, innovation is needed for new processes, products, and tools (Roco 2002). In fact, food nanotechnology introduces new opportunities for innovation in the food industry at immense speed. Some of the applications result in the presence of nanoparticles or nanostructured materials in the food. Nanotechnology has been revolutionizing the entire food system from production to processing, storage, and development of innovative materials, products, and applications. It could generate innovation in the macroscale characteristics of food, such as texture, taste, other sensory attributes, coloring strength, processability, and stability during shelf life, thus leading to a large number of new products. Moreover, nanotechnology can also improve the water solubility, thermal stability, and oral bioavailability of bioactive compounds (Huang et al. 2010; McClements et al. 2009; Silva et al. 2012; Ezhilarasi et al. 2013).

Currently, the market for nanotechnology products in the food industry approaches the US$ 1 billion (most of this on nanoparticle coatings for packaging applications, health promoting products, and beverages) and is predicted to rise to $20 billion in the next decade (Chau et al. 2007). At present, the main applications of nanotechnology in the food industry are nanocomposites in food packaging material (for controlling diffusion and microbial protection), nanobiosensors (for detection of contamination and quality deterioration), and nanoencapsulation or nanocarriers (for controlled delivery of nutraceuticals) (Chen et al. 2006a; Sanguansri and Augustin 2006; Sozer and Kokini 2009; Weiss et al. 2006; Ezhilarasi et al. 2013). Nanotechnology helps to make interactive foods that can allow consumers to modify the food depending on their own nutritional needs or tastes. The concept of on-demand food states that thousands of nanocapsules containing flavor or color enhancers or added nutritional elements remain dormant in the food and are only released when triggered by the consumer (Dunn 2004). Smaller particles improve food's spreadability and stability, and can aid in developing healthier low-fat food products.

1.2 Nanoencapsulation

Nanoencapsulation is defined as a technology to encapsulate substances in miniature and refers to bioactive packing at the nanoscale (Lopez et al. 2006). Encapsulation is a rapidly expanding technology with many potential applications

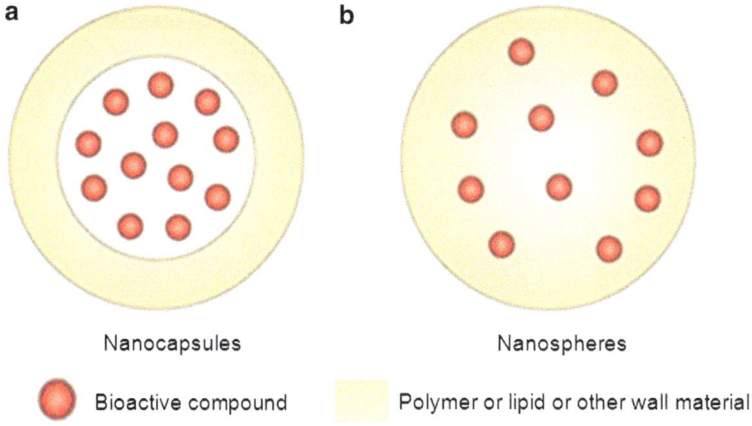

Fig. 1.1 Schematic structure of (**a**) nanocapsules and (**b**) nanospheres (Orive et al. 2009)

in areas such as the pharmaceutical and food industries. Encapsulation has been employed to protect bioactive compounds (polyphenols, micronutrients, enzymes, antioxidants, and nutraceuticals) and in the finished application, ultimately to protect them from adverse environmental factors and also for controlled release at the targeted site (Gouin 2004; Ezhilarasi et al. 2013).

A controlled and targeted release improves the effectiveness of micronutrients, broadens the application range of food ingredients, and ensures optimal dosage, thereby improving cost-effectiveness of the product (Mozafari et al. 2006). Microcapsules are particles having diameter between 1 and 5,000 μm (King 1995). Nanoparticles are colloidal-sized particles with diameters ranging from 10 to 1,000 nm and are expressed as nanocapsules or nanospheres (Konan et al. 2002). Nanocapsules are vesicular systems in which the bioactive compound is confined to a cavity surrounded by a unique polymer membrane, whereas nanospheres are matrix systems in which the bioactive compound is uniformly dispersed (see Fig. 1.1) (Couvreur et al.1995).

Nanoencapsulation involves the incorporation, absorption, or dispersion of bioactive compounds in the form of small vesicles of nanometer (or submicron) diameters (Taylor et al. 2005). The delivery of any bioactive compound to various sites within the body is directly affected by particle size (Kawashima 2001; Hughes 2005). Reducing the size of the encapsulates to the nanoscale offers opportunities related to prolonged gastrointestinal retention time due to improved bioadhesiveness in the mucus covering the intestinal epithelium (Chen et al. 2006b). Moreover, modulations of surface properties (e.g., coatings) can enable targeted delivery of compounds (Bouwmeester et al. 2009). Thus, nanoencapsulation has the potential to enhance bioavailability, improve controlled release, and enable precision targeting of the bioactive compounds to a greater extent than microencapsulation (Mozafari et al. 2006). Besides improving bioavailability, nanoencapsulation systems offer numerous benefits, including ease of handling, enhanced stability, protection against oxidation, retention of volatile ingredients, taste masking, moisture-triggered controlled release, pH-triggered controlled release, consecutive delivery of multiple active ingredients, change in flavor character, and long-lasting organoleptic perception (Shefer 2008; Ezhilarasi et al. 2013).

1.3   Materials for Nanoencapsulation

Nutraceuticals or bioactive compounds are extra nutritional compounds in the food that impart health benefits. Their effectiveness in preventing disease depends on preserving the bioavailability of the bioactive ingredient until it reaches the targeted site (Chen et al. 2006a). Reducing the particle size may improve the bioavailability, delivery properties, and solubility of the nutraceutical because of the greater surface area per unit volume and thus increased biological activity (Shegokar and Muller 2010). Bioavailability of the nutraceutical is increased because the nanocarrier allows it to enter the bloodstream from the gut more easily. The nanoscale nutraceutical was termed "nanoceutical" and the carrier was called a "nanocarrier" due to their size (Chen et al. 2006a). These nutraceutical compounds can be classified as lipophilic and hydrophilic on the basis of their solubility in water. Hydrophilic compounds are soluble in water but insoluble in lipids and organic solvents. Examples of nanoencapsulated hydrophilic nutraceuticals are ascorbic acid, polyphenols, etc. (Lakkis 2007; Teeranachaideekul et al. 2007; Dube et al. 2010; Ferreira et al. 2007). Lipophilic compounds are insoluble in water but soluble in lipids and organic solvents. Nanoencapsulated lipophilic nutraceuticals include lycopene, beta-carotene, lutein, phytosterols, and docosahexaenoic acid (DHA) (Lakkis 2007; Heyang et al. 2009; Zimet and Livney 2009; Leong et al. 2011). Solubility of the bioactive ingredients determines the release rate and release mechanism from a polymeric matrix system. Hydrophilic compounds show faster release rates and the release kinetics are determined by the appropriate combination of diffusion and erosion mechanisms. Lipophilic compounds often result in incomplete release due to poor solubility and low dissolution rates by an erosion mechanism (Kuang et al. 2010; Kumar and Kumar 2001; Varma et al. 2004). However, lipophilic compounds are highly permeable through the intestinal membrane via active transport and facilitated diffusion, whereas hydrophilic compounds have low permeability and are absorbed only by active transport mechanisms (Acosta 2009). Besides solubility, a bioactive compound needs to retain its stability and biological activity until it reaches the targeted site (Ezhilarasi et al. 2013).

Nanoencapsulation protects the bioactive component from unfavorable environmental conditions, e.g., from oxidation and pH and enzymatic degradation (Fang and Bhandari 2010). Nanocarrier food systems such as lipid-based or natural biodegradable polymer-based capsules are most often utilized for encapsulation (Chen et al. 2006a). Nanoemulsions, nanoliposomes, solid lipid nanoparticles, and nanostructured lipid carriers are widely used lipid-based nanocarrier systems having application in the pharmaceutical, cosmetic, and food industries. Natural polymers such as albumin, gelatin, alginate, collagen, chitosan, and α-lactalbumin are used in the formulation of nanodelivery systems (Reis et al. 2006; Graveland-Bikker and De Kruif 2006). There has been tremendous growth in the development of food nanocarrier systems in the last decade as various products have been developed (e.g., whey protein) for use as nanocarriers to improve the bioavailability of nutraceuticals, for nanodrop mucosal delivery of vitamins, and as nanobased mineral delivery systems (Chen et al. 2006a, b; Ezhilarasi et al. 2013).

1.4   Nanoencapsulation Techniques

The techniques used for achieving nanoencapsulation are more complex than for microencapsulation. This is mainly due to the difficulty in attaining the complex morphology of the capsule and core material and the demands of controlling the release rate of the nanoencapsulates (Chi-Fai et al. 2007). The chemical and physical natures of the core and shell materials, evaluation of their interactions, as well as their proportion in the formulation of the capsules are important parameters that determine the final properties of the particles (Li and Szoka 2007). Various techniques have been developed and used for microencapsulation purposes. However, emulsification, coacervation, inclusion complexation, emulsification-solvent evaporation, nanoprecipitation, and supercritical fluid techniques are considered to be nanoencapsulation techniques because they can produce capsules in the nanometer range (10–1,000 nm) (Ezhilarasi et al. 2013). The nanoencapsulation techniques use either top-down or bottom-up approaches for the development of nanomaterials. Top-down approaches involve the application of precise tools with force that allow size reduction and shaping of the structure for the desired application of the nanomaterial. The degree of control and refinement in size reduction processes influences the properties of the materials produced. In the bottom-up approach, materials are constructed by self-assembly and self-organization of molecules, which are influenced by many factors, including pH, temperature, concentration, and ionic strength (Augustin and Sanguansri 2009).

Techniques such as emulsification and emulsification-solvent evaporation are used in the top-down approach, On the other hand, supercritical fluid techniques, inclusion complexation, coacervation, and nanoprecipitation are used in the bottom-up approach (as shown in Fig. 1.2) (Sanguansri and Augustin 2006; Mishra et al. 2010). These nanoencapsulation techniques can be used for encapsulation of various hydrophilic and lipophilic bioactive compounds. Emulsification, coacervation, and the supercritical fluid technique are used for encapsulation of both hydrophilic and lipophilic compounds (McClements et al. 2009; Chong et al. 2009; Leong et al. 2009). However, inclusion complexation, emulsification-solvent evaporation, and nanoprecipitation techniques are mostly used for lipophilic compounds (Reis et al. 2006; Ezhilarasi et al. 2013).

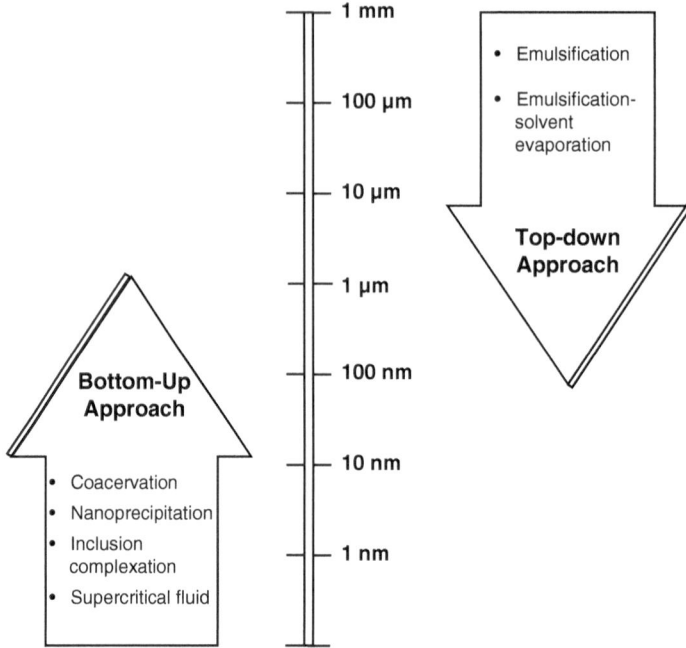

Fig. 1.2 Top-down and bottom-up approaches in nanoencapsulation (Ezhilarasi et al. 2013)

For understanding the possible benefits as well as the potential toxicity of nanoparticles in biological systems, complete and accurate characterization of nanoparticles is required (Oberdorster et al. 2005). Characterization includes state of aggregation, dispersion, sorption, size, structure, and shape, which can be studied using transmission electron microscopy (TEM), scanning electron microscopy (SEM), high-resolution transmission electron microscopy (HRETM), and atomic force microscopy (AFM) (Mavrocordatos et al. 2004). Tiede et al. (2008) reviewed the measurement of nanoparticle properties using different analytical techniques such as microscopy, chromatography, spectroscopy, centrifugation, and filtration.

The main focus of subsequent chapters is to discuss the various nanoencapsulation techniques, especially for food applications, along with the safety and regulatory issues.

Chapter 2
Techniques for Formation of Nanoemulsions

A combination of high-energy approaches (such as high-speed and high-pressure homogenization or high-pressure homogenization and ultrasonication) can aid in the formation of nanoemulsions with very small droplet diameters. A practical approach is to emulsify the sample under conditions of increasing intensity (e.g., starting at 2,000 rpm and increasing to 20,000 rpm in a rotor-stator), especially if the dispersed phase is highly viscous. One major constraint faced by researchers after the production of nanoemulsions is the process of Ostwald ripening, wherein the mean size of the nanoemulsion increases over time due to diffusion of oil molecules from the small to large droplets through the continuous phase. This is particularly seen in nanoemulsions formed by low-energy emulsification methods. A possible means to overcome this instability mechanism is by increasing the surfactant concentration by altering the oil-to-surfactant ratio. This chapter reviews the various techniques for nanoemulsion preparation.

2.1 Nanoemulsions

Nanoemulsions are colloidal dispersions comprising two immiscible liquids, one of which is dispersed in the other, possessing droplets of diameter ranging from 50 to 1000 nm (Sanguansri and Augustin 2006). The very small droplet size provides nanoemulsion stability against sedimentation and creaming, along with a transparent or slightly turbid appearance suitable for food applications (Tadros et al. 2004). Based on the relative spatial organization of the oil and aqueous phases, there are two basic types of nanoemulsion: oil-in-water (O/W) and water-in-oil (W/O). The O/W nanoemulsion consists of oil droplets dispersed in a water phase and the W/O nanoemulsion consists of water droplets dispersed in an oil phase (McClements and Rao 2011).

Nanoemulsions offer great potential to encapsulate high concentrations of oil-soluble bioactive compounds into a wide range of foodstuffs. Lipophilic active

C. Anandharamakrishnan, *Techniques for Nanoencapsulation of Food Ingredients*, SpringerBriefs in Food, Health, and Nutrition, DOI 10.1007/978-1-4614-9387-7_2, © C. Anandharamakrishnan 2014

agents such as β-carotene, plant sterols, carotenoids, and essential fatty acids can be encapsulated and delivered by O/W emulsion, whereas a W/O emulsion can be used to encapsulate water-soluble food active agents such as polyphenols (Zuidam and Shimoni 2010). In order to protect these bioactive compounds from degradation, a stable nanoemulsion has to be obtained. Hence, forming a nanoemulsion possessing a size specific for the end application proves to be a crucial process. However, the systems are non-equilibrium, they cannot be formed spontaneously and require energy input provided by external or internal sources (Gutierrez et al. 2008; Meleson et al. 2004). Besides the source of energy, a few other factors affect the final stability and characteristics of the nanoemulsion. For instance, the method of preparation, the order of addition of the ingredients (e.g., oil, water, and surfactant), the nature of the continuous and dispersed phases, and the amount and type of surfactant used, together affect the formation of nanoemulsions (Tadros et al. 2004; Devarajan and Ravichandran 2011).

The fabrication of nanoemulsions can be broadly categorized as either high-energy or low-energy approaches, depending on the underlying principle. The high-energy approaches disrupt the oil and aqueous phases into tiny droplets using mechanical devices such as high-pressure homogenizers, microfluidizers, and soni-cators (Leong et al. 2009). In low-energy approaches, nanoemulsions are formed as a result of phase transitions that occur during the emulsification process when the environmental conditions (either temperature or composition) are altered, e.g., phase inversion and spontaneous emulsification methods (Yin et al. 2009). Table 2.1 describes the nanoemulsification techniques for various bioactive components.

2.2 High-Energy Approaches

In high-energy approaches, intense disruptive force is applied on the sample to be emulsified. Two opposing processes namely, droplet disruption and droplet coales-cence, take place inside the system. The achievement of a balance between these two processes leads to the production of smaller droplets (Jafari et al. 2008). Intense energies are required from a mechanical device or homogenizer to generate disrup-tive forces higher than the restoring forces that hold the droplets into spherical shapes (Schubert and Engel 2004). The droplet size of nanoemulsions depends on factors such as the design of homogenizer (rotor-stator homogenizer, pressure homogenizer, ultrasonic homogenizer), homogenizer operating conditions (pres-sure, temperature, number of passes or cycles, valve and impingement design, flow rate), environmental conditions (temperature), sample composition (oil phase, aqueous phase, surfactant concentration, and/or co-surfactant concentration), and its physicochemical properties (Kentish et al. 2008; Wooster et al. 2008; Bhavsar 2011). During emulsification, an increase in the energy intensity or duration will decrease the interfacial tension. This enhances the emulsifier adsorption rate and, consequently, the disperse-to-continuous phase viscosity ratio (η_D/η_C) falls within a certain range, i.e., 0.05–5 (Tadros et al. 2004). These processes collectively aid in

Table 2.1 Nanoemulsification techniques for various bioactive compounds

Nanoemulsification technique(s)	Emulsion system (oil in water)	Principal compound(s)	Mean droplet diameter	Purposes	References
High-pressure homogenization (600 bar, 6 cycles)	Sodium dihydrogen-phosphate and protein stearin-rich milk fat fraction	α-Tocopherol	120 nm to 120 μm	Protection against degradation	Relkin et al. (2008)
High-pressure homogenization (1,500 bar, 6 cycles)	MCT and Tween 20	Curcumin	82 nm	Enhanced anti-inflammatory activity	Wang et al. (2008b)
High-pressure homogenization (300 MPa, 10 cycles)	Palm oil, sunflower oil, soy lecithin, clear gum, Tween 20, and glycerol monooleate	Terpene mixture and D-limonene	Terpenes, 75 nm; D-limonene, 240 nm	Enhanced antimicrobial activity	Donsi et al. (2011b)
High-pressure homogenization	MCT oil; Tween 20, 40, 60, and 80	β-Carotene	132–184 nm	Increased stability	Yuan et al. (2008)
Microfluidization (60 MPa)	Tween 80	Lemon myrtle oil	95–99 nm	Stable emulsions with pleasant mild flavor	Buranasuksombat et al. (2011)
Microfluidization (9,000 psi, three passes)	Orange oil and β-lactoglobulin	β-Carotene	156 nm	Enhanced chemical stability	Qian et al. (2012)
Microfluidization (150 MPa, three passes)	Sodium alginate and Tween 80	Lemongrass oil	7 nm	Properties of nanoemulsions changed with operating conditions	Salvia-Trujillo et al. (2013)
Ultrasonication (batch and continuous focused flow)	Tween 40	Flax seed oil	135 nm	Optimized operating conditions to prevent coalescence and cavitational bubble cloud formation	Kentish et al. (2008)

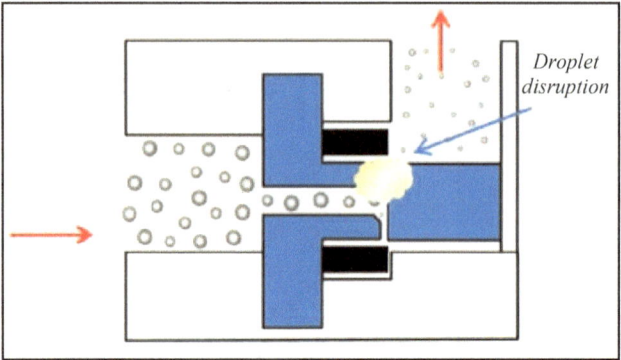

Fig. 2.1 High-pressure valve homogenizer (McClements and Rao 2011)

producing smaller droplet sizes. Preparation of a nanoemulsion is most efficiently carried out in two steps: first, conversion of separate oil and water phases into a "coarse emulsion" with fairly large emulsion droplet size (EDS) using rotor-stator devices (Rodgers et al. 2011) and, second, reduction of EDS using high-pressure systems (Jafari et al. 2008).

2.2.1 High-Pressure Homogenizer

High-pressure valve homogenizers are commonly used in the food industry for the production of conventional emulsions with small droplet sizes (Schubert et al. 2003; Schubert and Engel 2004). In nanoemulsion production, the coarse emulsions produced using rotor-stator systems are fed directly into the inlet of a high-pressure valve homogenizer. Then, the emulsion is pulled into a chamber by the backstroke of a pump and forced through a narrow valve present at the end of the chamber on its forward stroke (Tadros et al. 2004). The passage of the coarse emulsion through the narrow valve causes the large droplets to be broken down into smaller ones by a combination of intensive disruptive forces acting on them, such as shear stress, cavitation, and turbulent flow conditions (Fig. 2.1) (Stang et al. 2001).

Generally, it requires extremely high pressure and multiple passes to produce nanoemulsions with the required EDS (McClements and Rao 2011). Other factors that determine the droplet size are the disperse-to-continuous phase viscosity ratio (η_D/η_C) and the usage of an appropriate emulsifier (Walstra 1993, 2003). Using high-pressure homogenization, Yuan et al. (2008) produced β-carotene nanoemulsions (O/W) (Fig. 2.2) with medium chain triacylglycerol (MCT) oil. Various types of emulsifiers (Tween 20, 40, 60, and 80) at various concentrations from 4% to 12% were used along with changes in homogenization pressure, temperature, and cycle. Nanoemulsions with dispersed particle sizes ranging from 132 to 184 nm were obtained. The particle sizes decreased and physical stability increased with increase in homogenization pressure and cycle.

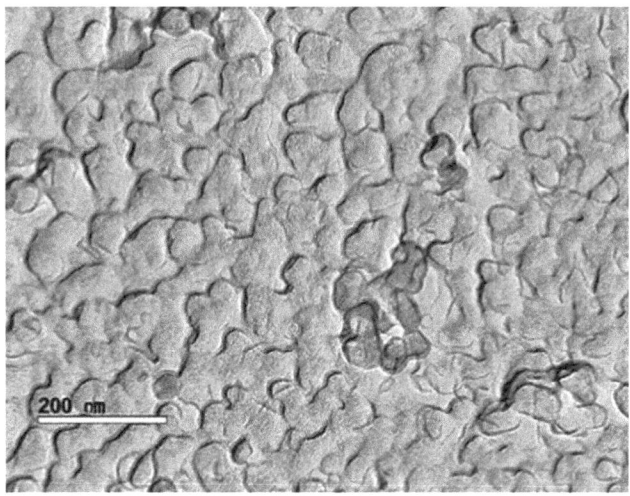

Fig. 2.2   Transmission electron micrograph of β-carotene nanoemulsions (Yuan et al. 2008)

Similarly, Wang et al. (2008b) prepared curcumin nanoemulsions using MCT as oil and Tween 20 as emulsifier by high-pressure homogenization. Nanoemulsions had a mean droplet size of 82 nm with enhanced anti-inflammatory activity, as evidenced by experiments using the mouse ear inflammation model. Likewise, using high-pressure homogenization, Relkin et al. (2008) studied the structural behavior of lipid droplets (composed of stearin-rich milk fat fraction) in sodium caseinate-stabilized nanoemulsions containing α-tocopherol. Study revealed that the food-grade nanoemulsions effectively protected α-tocopherol against degradation. Recently, Donsi et al. (2011b) prepared nanoemulsions using a terpene mixture and D-limonene using high-pressure homogenization at 300 MPa. Various formulations were prepared using palm oil and sunflower oil as lipophilic phase and soy lecithin, clear gum, Tween 20, and glycerol monooleate as emulsifiers. The most promising formulations had mean droplet size of 75 nm for terpenes and 240 nm for D-limonene, along with enhanced antimicrobial activity in foods.

2.2.2   Microfluidizer

Microfluidization involves the application of high pressure on a coarse emulsion for the production of nanoemulsions. It produces smaller droplets and narrower distribution of EDS compared with traditional emulsification techniques (Pinnamaneni et al. 2003).

A microfluidizer is similar to a high-pressure homogenizer; however, the design of the channels for the flow of emulsion is different. The emulsion initially flowing through a channel is further divided into two streams and each stream is passed

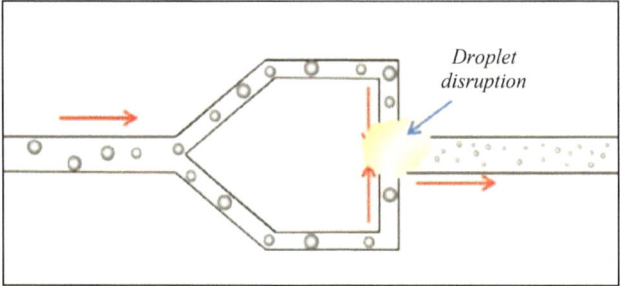

Fig. 2.3 Microfluidizer (McClements and Rao 2011)

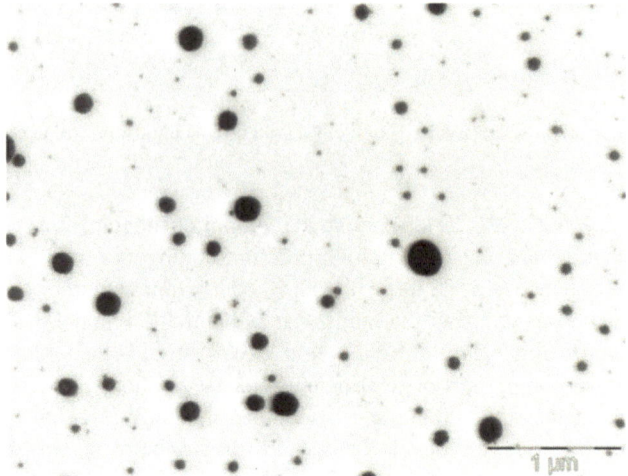

Fig. 2.4 Transmission electron micrograph of microfluidized lemongrass oil–alginate nanoemulsion (Salvia-Trujillo et al. 2013)

through a separate fine channel. At the interaction chamber, the two fast-moving streams are directed at each other, creating intense disruptive forces that lead to highly efficient droplet disruption (McClements and Rao 2011) (as shown in Fig. 2.3).

Using a microfluidization technique, Buranasuksombat et al. (2011) produced nanoemulsions of lemon myrtle oil with Tween 80 as emulsifier at 60 MPa pressure. The nanoemulsion had a mean droplet size of 95–99 nm, with a mild lemon flavor. Similarly, Qian et al. (2012) prepared β-carotene-enriched nanoemulsions stabilized by β-lactoglobulin using a high-pressure microfluidizer. A nanoemulsion with mean droplet diameter of 156 nm was obtained. The results also showed that the nano-emulsions were stable for 15 days at 20 °C, as evident by the unvarying particle radius and particle size distribution.

Recently, Salvia-Trujillo et al. (2013) studied the effect of processing parameters on the physicochemical characteristics of microfluidized lemongrass essential oil–alginate nanoemulsions (Fig. 2.4). The microfluidizer processing pressure and

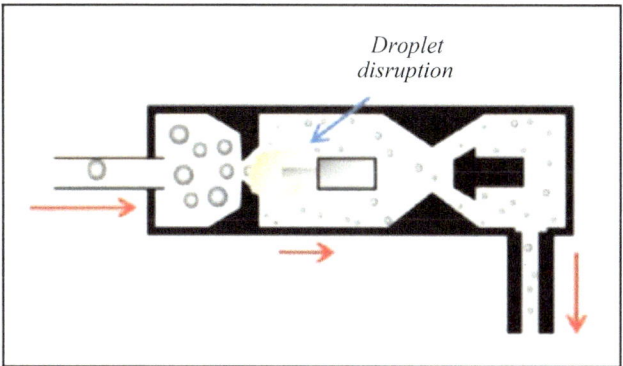

Fig. 2.5   Ultrasonic jet homogenizer (McClements and Rao 2011)

number of cycles significantly influenced the average droplet size, polydispersity index, and size distribution of the emulsion. When passed three times through a microfluidizer operating at 150 MPa the obtained nanoemulsions were almost transparent, with an average droplet size of 7 nm and polydispersity index of 0.34. The authors reported that microfluidization was an effective technique for formation of essential oil nanoemulsions.

Under certain conditions, such as high pressure and longer emulsification times, microfluidization is considered to be an unfavorable technique because it leads to "over-processing," which is re-coalescence of emulsion droplets (Jafari et al. 2006; Lobo and Svereika 2003; Olson et al. 2004). This phenomenon was also reported by Jafari et al. (2007a) in the preparation of D-limonene nanoemulsion (O/W) by microfluidization and ultrasonication and subsequent encapsulation using spray drying. Maltodextrin combined with a surface-active biopolymer (Hi-Cap) at a ratio of 3:1 was used as continuous phase. The study reported that microfluidization proved to be an efficient emulsification technique, however, increasing the microfluidization energy input beyond moderate pressures (40–60 MPa) and number of cycles (1–2) led to over-processing of emulsion droplets due to re-coalescence.

2.2.3   Ultrasonic Homogenizers

Ultrasonic emulsification involves the application of an acoustic field that produces interfacial waves to create intense disruptive forces, resulting in eruption of the oil phase into the water medium in the form of droplets (Li and Fogler 1978a).

The pressure fluctuations of the sound waves cause formation and subsequent collapse of microbubbles. This creates extreme levels of localized turbulence, which aid in breaking up primary droplets of dispersed oil into droplets of submicron size (Li and Fogler 1978b). Figure 2.5 illustrates the ultrasonic jet homogenizer. Ultrasound can be generated either mechanically (whistle, siren) or electrically (reverse piezoelectric effect or magnetostrictive transducers), the latter being widely used in food systems (Mason 1999). Emulsification can be achieved using ultrasound in the range of 16–100 kHz (Canselier et al. 2002), either as a batch or continuous process. Kentish et al. (2008) studied both a batch and continuous focused flow-through ultrasonic cell for emulsification using flax seed oil. Emulsions with a mean droplet size as low as 135 nm were achieved using Tween 40 as surfactant. The authors concluded that the batch cell produced better results although continuous equipment is likely to be more viable in a commercial environment.

2.3　Low-Energy Approaches

Low-energy methods use the internal chemical energy of the system for emulsion production (Tadros et al. 2004; Sole et al. 2006) and are more cost-effective than high-energy methods. Commonly used low-energy emulsification methods are self-emulsification and phase inversion methods.

2.3.1　Self-Emulsification Methods

Self-emulsification methods involve the spontaneous formation of emulsion through the rapid diffusion of surfactant and/or solvent molecules from the dispersed phase to the continuous phase with the help of the chemical energy released due to the dilution process (Pouton and Porter 2006; Ghai and Sinha 2012). This method can be used to produce nanoemulsions by dilution of microemulsions at stabilized mixing conditions. Dilutions can be brought about by addition of an organic phase containing hydrophobic oil, a hydrophilic surfactant, and a water-miscible organic solvent to water (Anton and Vandamme 2009), or by addition of water to an organic phase containing hydrophobic oil, water-miscible organic solvent, and surfactant (Sonneville-Aubrun et al. 2009). For instance, Porras et al. (2008) formed nanoemulsions with droplet size ranging from 30 to 120 nm by employing the low-energy emulsification method of adding water to a mixture of surfactants and decane (oil phase). In this way, when surfactant, oil, and water contents are optimized, nanoemulsions can be produced (McClements and Rao 2011).

2.3.2　Phase Inversion Methods

Phase inversion methods utilize the chemical energy released as a result of phase transitions taking place during the emulsification process. Nanoemulsions have been formed by inducing phase inversion in emulsions from a W/O to O/W form or vice versa by either changing the temperature (phase inversion temperature methods, PIT) or the composition (phase inversion composition methods, PIC) (Solans and Sole 2012; McClements and Rao 2011). The phase transition in PIT methods is achieved by changes in temperature that affect the spontaneous curvature of surfactants such as polyoxyethylene-type nonionic surfactants. But in the PIC method, phase transition is achieved by changes in the composition at a constant temperature (Solans and Sole 2012).

2.3.2.1 Phase Inversion Temperature Methods

The PIT method is based on the changes caused in the physicochemical properties of nonionic surfactants with a change in temperature (Anton and Vandamme 2009; Anton et al. 2008; Gutierrez et al. 2008). Manipulation of the temperature–time profile of certain mixtures of oil, water, and nonionic surfactant can aid in the formation of nanoemulsions using PIT methods.

The preparation of nanoemulsions using PIT methods requires the sample to be brought to its phase inversion temperature (PIT) or hydrophile–lipophile balance (HLB) temperature. This is the temperature at which the balance exists between hydrophilic and lipophilic properties of the system and, hence, an extremely low interfacial tension can be achieved to promote emulsification (Shinoda and Kunieda 1983; Kunieda and Friberg 1981). As a result, nanoemulsions can be obtained. However, at the HLB temperature, the barriers that oppose coalescence processes are low and this may lead to coalescence of the droplets, making the emulsion unstable (Taisne and Cabane 1998; Kabalnov and Wennerstrom 1996). Hence, once the small droplets are formed, the temperature has to be immediately shifted from the HLB temperature either by rapid cooling or by rapid heating. This essential step in the PIT method gives rise to kinetically stable nanoemulsions (Solans and Sole 2012).

2.3.2.2 Phase Inversion Composition Methods

The PIC method involves the progressive addition of one of the components either water or oil to a mixture of the other two components, either oil/surfactant or water/surfactant, respectively (Machado et al. 2012). As the volume fraction of the water increases in the system, a transition composition is obtained where the hydration grade of the polyoxyethylene chains of the surfactant progressively increases and the surfactant spontaneous curvature changes from negative to zero. In this transition composition, the surfactant hydrophilic–lipophilic properties are balanced (as obtained at the HLB temperature). Above the transition composition, the structures with zero curvature separate and form metastable O/W droplets, which have a very small diameter. The PIC method has potential advantages over the PIT method for handling components with temperature-stability problems or systems containing surfactants that are not of the polyoxyethylene type (Solans and Sole 2012).

2.3.3 Emulsion Inversion Point Methods

In PIT and PIC methods of emulsification, a change is brought about in the properties of the surfactant by manipulating the temperature, pH, or composition of the system, leading to a transitional-phase inversion from one type of emulsion to another. In the emulsion inversion point (EIP) method, catastrophic-phase inversion is induced by altering the ratio of the oil-to-water phases, either by increasing or decreasing the volume of the dispersed phase in an emulsion above or below some critical level (Fernandez et al. 2004; Thakur et al. 2008).

In the EIP method, an W/O emulsion containing water droplets dispersed in oil is formed initially using a small molecule surfactant. To achieve phase inversion from a W/O to O/W system, increasing amounts of water is added to the system with constant stirring until the critical point is exceeded. At this stage, the coalescence rate of water droplets exceeds the coalescence rate of oil droplets and leads to phase inversion (Thakur et al. 2008). The major factors contributing to the emulsification process are the critical surfactant concentration and the surfactant-to-oil ratio (Fernandez et al. 2004).

Nanoemulsions can be used directly in the liquid state or can be dried to powder form using drying techniques such as spray drying and freeze-drying. The application of food-grade nanoemulsions is discussed separately in Chap. 7.

Chapter 3
Bioactive Entrapment Using Lipid-Based Nanocarrier Technology

A delivery system is a formulation or a device that introduces bioactive compounds into the body and improves their efficacy and safety by controlling the rate and targeted release of the bioactive compound. Food-based nanocarrier systems are generally based on carbohydrate, protein, or lipids. There are various types of lipid-based nanoparticulate delivery systems such as nanoemulsions, nanoliposomes, solid lipid nanoparticles, lipid nanotubes, lipid nanospheres, and nanostructured lipid carriers. As nanocarriers, they are expected to be promising oral carriers due to their potential to improve the solubility and bioavailability of poorly water-soluble and lipophilic compounds. The selection of an appropriate delivery system is dictated by the characteristic solubility and stability of the bioactive compound, the safety and efficiency of the bioactive-carrier lipid matrix, intended application, route of administration, etc. Hence, lipid-based nanoencapsulation is emerging as one of the most promising encapsulation technologies in the field of nanotechnology. This chapter reviews the present state of the art of widely used lipid-based nanoparticulate delivery systems such as solid lipid nanoparticles (SLN), nanostructured lipid carriers (NLC), and nanoliposomes.

3.1 Lipid-Based Delivery Systems

A lipid-based delivery system has the ability to entrap material with different solubilities and can be produced using natural lipid based ingredients on an industrial scale (Mozafari and Mortazavi 2005). Such delivery systems can also protect the active compounds from biological degradation or transformation, enhance the bioactive compound potency with higher encapsulation efficiency, and can lower toxicity (Fathi et al. 2012). Another unique property of lipid-based nanocarriers is targeted delivery inside the body through active (e.g., by incorporation of antibodies) and passive mechanisms (e.g., based on particle size) (Mozafari and Mortazavi

C. Anandharamakrishnan, *Techniques for Nanoencapsulation of Food Ingredients*,
SpringerBriefs in Food, Health, and Nutrition, DOI 10.1007/978-1-4614-9387-7_3,
© C. Anandharamakrishnan 2014

2005). The delivery system can impart stability to water-soluble material and also accommodate lipid-soluble agents, with synergistic effect if required (Gouin 2004; Suntres and Shek 1996). Moreover, several studies have reported that encapsulation of bioactive compounds through lipid-based carrier systems improves their therapeutic potential by facilitating intracellular delivery and prolonging their retention time inside the cell (Stone and Smith 2004).

The major ingredients involved in the formulation of lipid-based particulate delivery systems are emulsifiers (e.g., Poloxamer 188, Polysorbate 20, lecithin), lipids (e.g., triglycerides, steroids), bilayer lipids (e.g., phospholipids) and non-bilayer lipids (e.g., hard fats, mono-, di-, and triglyceride mixtures) (Attama et al. 2012). These lipid nanoformulations can be tailored to meet a wide range of product requirements dictated by disease condition, route of administration, and considerations of cost, product stability, toxicity, and efficacy (Mozafari 2006). The reported safety and efficacy of lipid-based carriers make them attractive candidates for the encapsulation of nutraceuticals, pharmaceutical compounds, etc.

Lipid-based nanoparticulate delivery systems for various bioactive compounds are described in terms of their production method, advantages, flaws, and variations in Table 3.1.

3.2 Solid Lipid Nanoparticles

Solid lipid nanoparticles (SLN) are in the submicron size range and composed of solid lipids such as triacylglycerol, waxes, and paraffins (Jenning et al. 2000). The structure is an aqueous colloidal dispersion within a matrix of solid lipid shell (Muller et al. 2000). SLN are a new generation of submicron-sized lipid emulsions where the liquid lipid (oil) is replaced by solid lipid. Compared to nanoemulsions and liposomes, SLN have some distinct advantages (Saupe and Rades 2006), which include:

- High encapsulation efficiency
- Avoidance of organic solvents
- Possibility of large scale production and sterilization
- Flexibility in controlling the release of encapsulated compounds due to the solid matrix
- Slower degradation rate and sustained release of bioactive compound
- Effective protection of bioactive compounds against chemical degradation

High-pressure homogenization (HPH) is a suitable method for the preparation of SLN. High-pressure homogenizers push a liquid with high pressure (100–2,000 bar) through a narrow gap (in the range of a few microns). The fluid accelerates over a very short distance with very high velocity (over 1,000 km/h). Very high shear stress and cavitation forces disrupt the particles down to the submicron range. Typical lipid contents are in the range of 5–10 % and even higher lipid concentrations (up to 40 %) can be homogenized to lipid nanodispersions (Lippacher et al. 2000).

Table 3.1 Lipid-based nanoparticulate delivery systems for various bioactive compounds

Lipid based delivery system	Lipid compound and other materials used	Core compound	Particle size	Results	References
Solid lipid nanoparticles	Glyceryl behenate, hydrogenated soya bean, and lecithin	Resveratrol	<180 nm	Improved solubility, stability, and intracellular delivery	Teska and Kristl (2010)
	Hydrogenated palm oil	BSA	674–61 nm	Low in vitro cytotoxicity	Schubert and Muller-Goymann (2005)
	Stearic acid and GMS	Curcuminoids	450 nm	Prolonged release and increased storage stability	Tiyaboonchai et al. (2007)
	Trimyristin, tristerin, and GMS	Curcuminoids	120 and 250 nm	Higher entrapment efficiency	Nayak et al. (2010)
	N-Carboxymethyl chitosan, monoglyceride, and soya lecithin	Carvedilol	61–105 nm	Prevented burst release and protected from gastric environment	Venishetty et al. (2012)
	Stearic acid and soy lecithin	Curcuminoids	119 nm	Improved bioavailability	Li et al. (2011)
	Sodium deoxycholate	Quercetin and resveratrol	149 nm	Higher entrapment efficiency and stability	Cadena et al. (2013)
Liposomes	Phospholipids	Carnosine	116–120 nm	Morphology and encapsulation efficiency differ with the various phospholipids	Maherani et al. (2012)
	Soybean phospholipids	DHA and EPA	50–200 nm	Improved physical and oxidative stability	Rasti et al. (2012)
	Phytosterols and soy phospholipids	Ascorbic acid	115–150 nm	Improved stability	Alexander et al. (2012)
	Polyethylene glycol	Vitamin E	164 nm	Improved stability	Zhao et al. (2011)
	Phospholipid/coenzyme Q$_{10}$/cholesterol/Tween 80	Coenzyme Q$_{10}$	68 nm	Optimized the formulation	Xia et al. (2006)
	Phospholipon 90H and 100H, dipalmitoyl phosphatidylcholine, stearylamine, dicetyl phosphate, and cholesterol	Nisin	350 nm	Long-term stability	Colas et al. (2007)
	Phosphatidylcholine and cholesterol	Vitamin C	96 and 97 nm	Improved storage stability	Liu and Park (2010)

(continued)

Table 3.1 (continued)

Lipid based delivery system	Lipid compound and other materials used	Core compound	Particle size	Results	References
Nanostructured lipid carriers	Glycerol monolaurate, acetylated mono- and diglycerides, octyl and decyl glycerate, polyglyceryl-10 laurate, and soybean lecithin	Coenzyme Q_{10}	96 nm	Improved bioavailability and stability	Liu et al. (2012b)
	GMS, apricot kernel oil, PEG-6 ester, PEG-2 stearate, and nonionic hydrophilic white beeswax	Ascorbyl palmitate	<350 nm	Enhanced chemical stability	Teeranachaideekul et al. (2007)
	Sunflower oil and propylene glycol monostearate	β-Carotene	144–249 nm	Improved stability	Hentschel et al. (2008)
	Myverol, Pluronic, lauric acid, myristic acid, palmitic acid, stearic acid, and Precirol ATO5	Lutein	228–130 nm	Sustained-release	Liu and Wu (2010)

BSA bovine serum albumin, *DHA* docosahexaenoic acid, *EPA* eicosapentaenoic acid, *GMS* glyceryl monostearate, *PEG* polyethyleneglycol

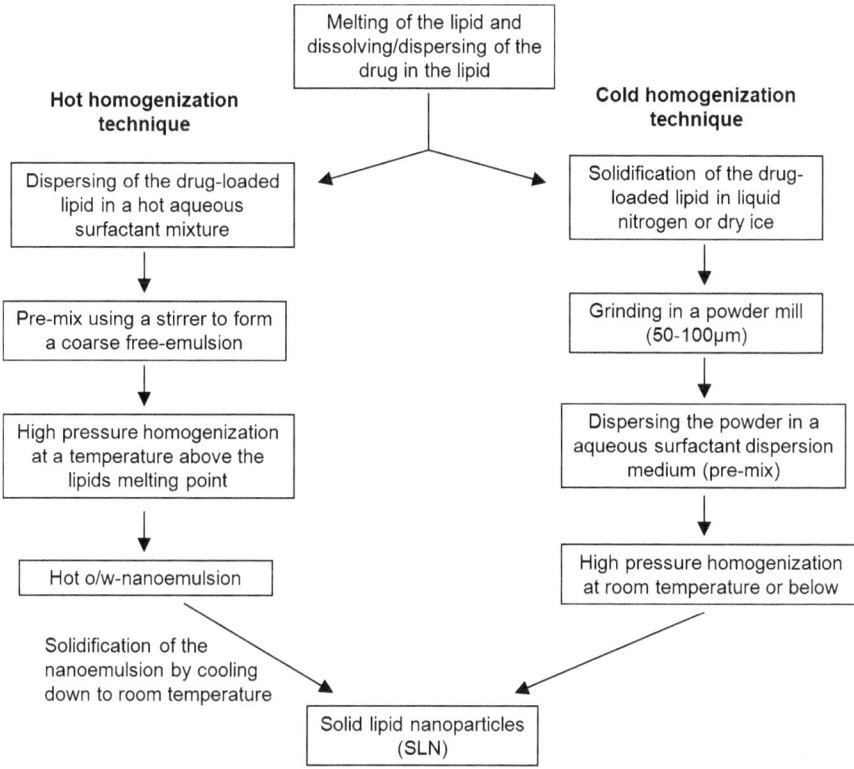

Fig. 3.1 Procedure for hot and cold homogenization technique in SLN production (Mehnert and Mader 2012)

In the production of SLN there are two general approaches for the homogenization step: hot and cold homogenization techniques (as shown in Fig. 3.1) (Muller et al. 1995). In both cases, a preparatory step involves the incorporation of active compounds into the bulk lipid by dissolving or dispersing them in the lipid melt technique. Briefly, for the hot HPH, the lipid and active compounds are melted (approximately 10 °C above the melting point of the lipid) and combined with an aqueous surfactant solution at the same temperature. A hot pre-emulsion is formed by homogenization and then processed in a temperature-controlled high-pressure homogenizer at 500 bar (or more) with a predetermined number of cycles. The obtained nanoemulsion recrystallizes upon cooling down to room temperature, forming SLN and nanostructured lipid carrier (NLC) nanoparticles. Recrystallization can also be initiated by lyophilization (Muller et al. 2000). The hot HPH is also suitable for heat-sensitive compounds because the exposure to an increased temperature is relatively short.

The cold HPH is a suitable technique for processing heat-labile or hydrophilic compounds. Here, lipid and active compounds are melted together and then rapidly ground under liquid nitrogen, forming solid lipid microparticles. This is then homogenized in a cold surfactant solution and subjected to HPH below room temperature under predetermined homogenization conditions to produce SLN and NLC nanoparticles (Lippacher et al. 2002).

Schubert and Muller-Goymann (2005) encapsulated BSA in solid lipid nanoparticles using hydrogenated palm oil using a HPH technique. The incorporation of lecithin up to 30 % (w/w) within the lipid matrices led to a concentration-dependent particle size reduction down to 100 nm. Variations in SLN composition resulted in particle sizes between 674 and 61 nm. The particles were anisometrical and crystalline with an albumin payload ranging from 2.5 % to 15 %. They also showed low in vitro cytotoxicity. Similarly, using high-speed homogenization through a melt emulsification process, Teska and Kristl (2010) encapsulated resveratrol in SLN. The solubility, stability, and intracellular delivery of resveratrol were increased by loading into SLN and the particle size was below 180 nm. The release profile of resveratrol showed a biphasic pattern and effectively decreased cell proliferation.

Microemulsion is another method for the preparation of SLN and it is based on the principle that addition of a microemulsion to water leads to precipitation of the lipid phase, forming particles (Gasco 1997). The excess water is then removed either by ultrafiltration or by lyophilization in order to increase the particle concentration. Using the microemulsion technique, Tiyaboonchai et al. (2007) developed curcuminoid-loaded SLN at 75 °C. At optimized process conditions, lyophilized curcuminoid-loaded SLN exhibited a spherical shape with a mean particle size of 450 nm and incorporation efficacy of 70 %. In vitro release studies showed a prolonged release of the curcuminoids from the SLN for up to 12 h. Moreover, lyophilized curcuminoids showed good physical and chemical stability over the storage period of 6 months.

SLN can also be prepared by ultrasonication techniques in combination with high-speed homogenization. Nayak et al. (2010) produced curcuminoid-loaded SLN by an emulsion technique employing a high-speed homogenizer and ultrasonic probe. Particles exhibited a spherical shape with sizes ranging between 120 and 250 nm. The entrapment efficiency and loading capacity was found to be in the range of 80–94 % and 1–3 %, respectively. Curcuminoid-loaded SLN exhibited a two fold increase in antimalarial activity as compared with free curcuminoids at the tested dosage level. Likewise, Li et al. (2011) prepared curcuminoid-loaded SLN and curcumin-loaded SLN with stearic acid and soy lecithin by using ultrasonication. The average particle size was 119 nm and the extraction recovery rate ranged from 78 % to 88 %. Recently, Venishetty et al. (2012) prepared N-carboxymethyl chitosan-coated carvedilol-loaded SLN through high-speed homogenization and ultrasonication techniques using monoglyceride as lipid. Particle size was in the range of 61–105 nm and encapsulation efficiency was about 95–98 %. Even after 3 months of storage, particle size remained below 100 nm. There was no burst release of carvedilol and it was effectively protected from the gastric environment, with improved stability in acidic medium.

3.3 Nanoliposomes

Liposomes are microscopic vesicles composed of membrane-like phospholipid bilayers surrounding an aqueous medium. Liposomes are successfully employed for the encapsulation of a wide range of synthetic and biological compounds with extensive application in the pharmaceutical, food, and cosmetics industries. Owing to the possession of both lipid and aqueous phases, liposomes can be utilized in the entrapment of water-soluble, lipid-soluble, and amphiphilic materials. Moreover, they are efficient at encapsulating and stabilizing bioactive molecules against various environmental factors and protecting them from degradation (Mozafari and Khosravi-Darani 2007).

Liposomes are classified according to the number of lamellae and size (Fig. 3.2). Liposomes with a single bilayer membrane are called small (<30 nm) or large (30–100 nm) unilamellar vesicles (ULV). Liposomes composed of a number of concentric bilayers are known as a multilamellar vesicles (MLV), and those composed of many small nonconcentric vesicles within a single lipid bilayer are known as a multivesicular vesicles (MVV) (New 1990). MLVs and MVVs are more suitable for entrapment of lipid-soluble material, whereas ULVs are suitable for water-soluble material. Apart from the core material, liposome formation is also based on the hydrophobic–hydrophilic interactions between lipid/lipid and lipid/water phases (Goyal et al. 2005).

Nanoliposomes are nanometric versions of liposomes, mostly applied for encapsulation and controlled release systems. Liposomes and nanoliposomes that can incorporate and release two materials with different solubilities simultaneously are termed "bifunctional liposomes" (Suntres and Shek 1996). Nanoliposomes should be capable of retaining their nanometric size and stability throughout the storage period and also during the end application (Mozafari and Mortazavi 2005). An input of energy results in arrangement of the lipid molecules, in the form of bilayer vesicles, to achieve a thermodynamic equilibrium in the aqueous phase.

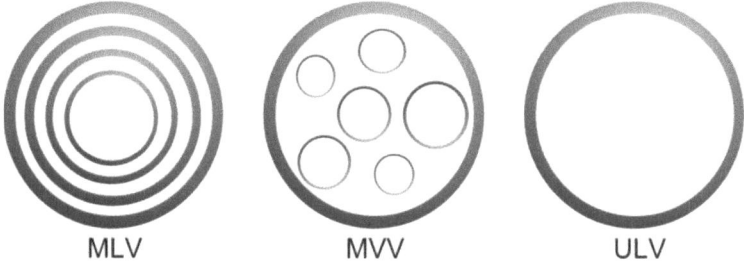

Fig. 3.2 Multilamellar vesicle (*MLV*), multivesicular vesicle (*MVV*), and unilamellar vesicle (*ULV*). The *shaded areas* are the lipidic phases of the liposomes, while the *enclosed white areas* are their aqueous phases (Mozafari et al. 2006)

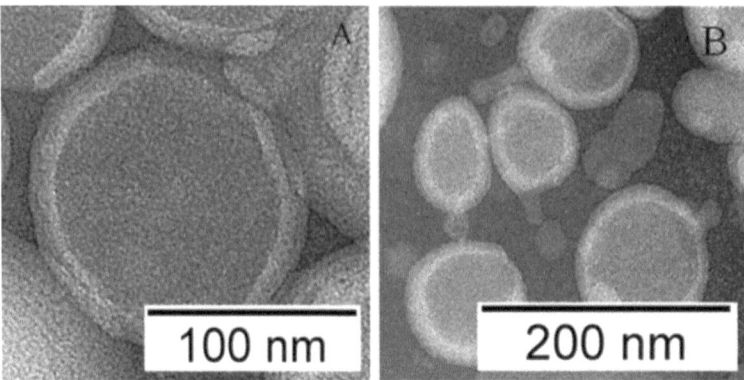

Fig. 3.3 Transmission electron micrographs of L-carnosine-encapsulated nanoliposomes prepared by the extrusion technique (Maherani et al. 2012)

The most commonly applied techniques based on the input of mechanical energy are high-intensity ultrasonication, high-pressure homogenization, extrusion, membrane homogenization, and microfluidization. Non-mechanical methods include reverse phase evaporation, removal of detergents from mixed detergent/lipid micelles, and freeze-drying followed by rehydration (Taylor et al. 2005).

Bangham thin film hydration is one of the most widely used methods for the preparation of MLV. The technique involves drying a solution of lipids so that a thin film is formed at the bottom of a round-bottom flask, followed by hydrating the film by adding aqueous buffer and vortexing the dispersion for some time. The compounds to be encapsulated are added either to aqueous buffer or to an organic solvent containing lipids, depending upon their solubilities (Bangham et al. 1965). Extrusion is another process in which micrometric liposomes (e.g., MLV) are structurally modified to large unilamellar vesicles (LUV) or nanoliposomes, depending on the pore size of the filters used (Berger et al. 2001).

Using a thin film hydration method together with extrusion, Maherani et al. (2012) encapsulated carnosine in different phospholipids by formation of nanoliposomes (Fig. 3.3). The particle size was in the range of 116–120 nm and loading efficiency was less than 22 %. The encapsulation efficiency was less than 25 % and it was determined that encapsulation efficiency tended to increase as the saturation degree of the lipids in the liposome membrane increases. Similarly, Cadena et al. (2013) performed nanoencapsulation of quercetin and resveratrol into sodium deoxycholate-elastic liposomes using a thin lipid film method along with freeze-drying. The obtained liposomes exhibited a mean diameter of 149 nm and the encapsulation efficiency was almost 97 %. The best liposomal formulation reduced the use of phosphatidyl choline and cholesterol to 18 % and 69 %, respectively.

Mozafari's heating method is a solvent-free method in which glycerol is used as dispersant agent, which facilitates the homogenous solubilization of lipids and encapsulated materials. The liposomal ingredients are added to a preheated mixture of bioactive compound and glycerol (at about 60 °C). The mixture is further heated while stirring under nitrogen atmosphere and results in the formation of MLV. If nanosized vesicles (nanoliposomes) are required, the samples are then extruded through membrane filters above the phase transition temperature (T_c) of the liposomes. Finally, the product is left at a temperature above T_c under nitrogen for stabilization (Mozafari 2005). Using this technique, Colas et al. (2007) encapsulated nisin and reported 12–54 % encapsulation efficiency. Moreover, the nanoliposomes were stable for 14 months at 4 °C and 12 months at 25 °C. Recently, Rasti et al. (2012) compared Bangham thin-film hydration and Mozafari methods by loading polyunsaturated fatty acids (PUFAs, docosahexaenoic acid and eicosapentaenoic acid) in liposomes using soybean phospholipids. The highest physicochemical stability was observed in PUFA liposomes prepared by the Mozafari method, rather than in those prepared by the Bangham method or in bulk PUFAs. Further, there was no significant change in physicochemical stability during 10 months of cold storage (4 °C) in the dark. Moreover, PUFAs in nanoliposomes (50–200 nm) exhibited higher surface charge, physical stability, and oxidative stability compared to those in liposomes (>200 nm).

HPH and ultrasonication can also be used to produce nanoliposomes. Using ethanol injection and sonication, Xia et al. (2006) produced coenzyme Q_{10} (CoQ10) nanoliposomes and optimized the formulation and production at the pilot scale. The best formulation CoQ10 nanoliposomes was found to be phospholipid/CoQ10/cholesterol/Tween 80 (2.5:1.2:0.4:1.8, w/w) with phosphate buffer solution as the hydration media. The average diameter was about 68 nm and encapsulation efficiency was greater than 95 % with a retention ratio higher than 90 %. The particle size changed by less than 10 % after storage at 4 °C in the dark for 90 days. Fluorescence probe studies indicated that CoQ10 incorporation increased the microviscosity of the nanoliposomes and inhibited the peroxidation of phospholipid. It was suggested that CoQ10 might intercalate between lipid molecules and perturb the bilayer structure. Similarly, Liu and Park (2010) developed vitamin-C-loaded chitosan-coated nanoliposomes using phosphatidylcholine (pc) and cholesterol (chol) by direct injection along with sonication. Liposomes prepared using ethanol as a solvent with pc: chol ratios of 40:60 and 60:40 exhibited mean diameters of about 97 and 96 nm, respectively. Liposomes prepared with a pc: chol ratio of 60:40 were promising vitamin C carriers with a maximum loading efficiency of about 96 % and payload of about 47 %. An increase in initial mass of vitamin C resulted in a higher payload. The liposomes prepared with 100 mg initial mass of vitamin C had maximum loading. Moreover, the loading efficiency and payload of liposomes were not affected by chitosan concentration. Liposomes prepared in the optimum conditions were stable during storage for 15 weeks and nearly 85 % of the vitamin C was protected against oxidation.

Zhao et al. (2011) produced conventional liposomes and PEG-coated vitamin E lyophilized proliposomes (PLP) by thin-film ultrasonic dispersion and lyophilization. The mean diameter and encapsulation efficiency for PEG-coated lyophilized proliposomes were 164 nm and 84 % respectively. The vitamin E contained within PLP exhibited a better stability than that in conventional liposomes and the retention percentage of PLP was 90 % at 4 °C after 15 days of storage. Recently, Alexander et al. (2012) incorporated phytosterols in soy phospholipids and encapsulated ascorbic acid using HPH. The plant sterols were incorporated at two different phospholipid/plant sterol mix ratios, 14:1 and 13:2 (g/g). All the obtained liposomes showed an initial monomodal size distribution with an average diameter of 115–150 nm. Incorporation of plant sterols increased the vesicle size and their encapsulation efficiency.

3.4 Nanostructured Lipid Carriers

Nanostructured lipid carriers (NLC) are colloidal carriers characterized by a solid lipid core consisting of a mixture of solid and liquid lipids with mean particle size in the nanometer range. They consist of a lipid matrix with a special nanostructure. NLC are considered to be second generation lipid nanoparticles (Muller et al. 2002) and can remain in the solid state by controlling the liquid lipid content added to the formulation. NLC also have the advantages of SLN, including low toxicity, biodegradation, protection, slow release, and avoidance of organic solvents during production (Han et al. 2008). Because of their good stability and high loading capacity, NLC are widely applied in the pharmaceutical field (Zhuang et al. 2010). With regard to these features, high drug payload, avoidance or minimization of drug expulsion, and enhancement of chemical stability can be achieved. Therefore, NLC have been described as a new generation of promising colloidal carriers.

Compared to emulsions, NLC can more strongly immobilize drugs and prevent the particles from coalescing by virtue of the solid matrix (Uner 2006). In addition, the mobility of the incorporated drug molecules is also drastically reduced in the solid phase. Furthermore, the liquid oil droplets in the solid matrix increase the drug payload capacity compared to SLN. NLC can be prepared either by blending any solid and liquid lipids or by mixing special combinations of solid and liquid lipids, leading to amorphous solids. Surfactants play an important role in the formation and characterization of NLC. Similar to SLN, the NLC can be manufactured through either hot or cold HPH. Other production methods for NLC are microemulsions (Gasco 1993), precipitation (Trotta et al. 2003), and dispersion by ultrasound (Domb 1993).

Teeranachaideekul et al. (2007) incorporated ascorbyl palmitate into nanostructured lipid carriers through a HPH method. The ascorbyl palmitate-loaded NLC exhibited enhanced chemical stability and storage stability. The storage of NLC at 4 °C and room temperature (25 °C) for 90 days retained more than 85 % of ascorbyl palmitate. Similarly, Hentschel et al. (2008) prepared β-carotene-loaded NLC by HPH using Tween 80 as emulsifier. The NLC had a mean diameter of 144–249 nm. The particle size decreased as the emulsifier concentration increased. All particles were smaller than 1 μm and had a mean particle size of around 0.3 μm during the storage period of 9 weeks at 20 °C and 30 weeks at 4–8 °C. Moreover, the addition of tocopherol improved the stability of β-carotene in the NLC.

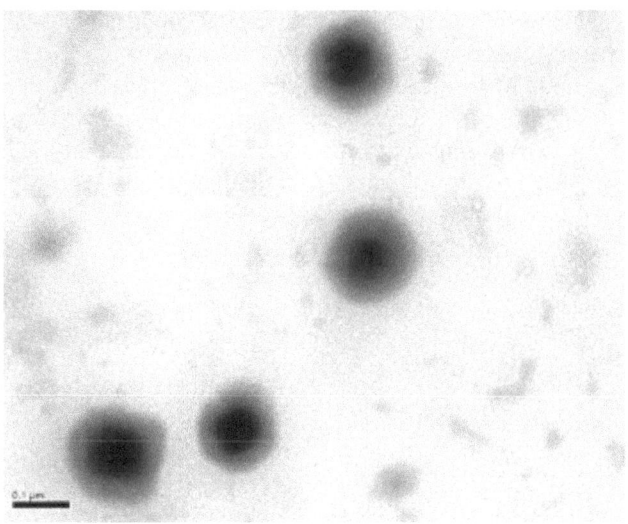

Fig. 3.4 Transmission electron micrographs of lutein-loaded NLC; *scale bar* 100 nm (Liu and Wu 2010)

Using ultrasonication, Liu and Wu (2010) encapsulated lutein in NLC and prepared a stable NLC dispersion using Precirol, Myverol, and Pluronic. The prepared NLC had an imperfect crystalline lattice and a spherical morphology (Fig. 3.4). The size of the NLC decreased from 228 to 130 nm as the time of ultrasonication increased from 2 to 10 min. Lutein-loaded NLC effectively protected lutein in simulated gastric fluid, and slowly released lutein in simulated intestinal fluid. Recently, Liu et al. (2012b) produced coenzyme Q_{10}-loaded NLC by hot HPH along with high-speed stirring at 10,000 rpm. The coenzyme Q_{10}-loaded NLC showed an ellipsoid morphology and average diameter of about 96 nm. They had average encapsulation efficiency of about 98 % and exhibited good physical and chemical stability. After lyophilization, the NLC had good storage stability and exhibited high bioavailability in the in vivo study.

Recently, Aditya et al. (2013) fabricated curcumin- and genistein-loaded NLC and evaluated the impact of the carriers on the bioaccessibility of curcumin and genistein. The entrapment efficiency was more than 75 % for curcumin- and/or genistein-loaded NLC. Their solubility in simulated intestinal medium was greater than 75 %, compared with less than 20 % for unencapsulated compounds. Both curcumin and genistein shown good stability (greater than 85 %) in simulated intestinal and gastric medium for up to 6 h. Co-loading of curcumin and genistein had no adverse effect on the solubility and stability of each molecule. Instead, co-loading increased loading efficiency and cell growth inhibition in prostate cancer cells.

Lipid-based nanoparticulate delivery systems have been proved to be one of the best platforms for enhancing the stability, oral bioavailability, and biological efficacies of different bioactive compounds. They are especially appealing to the food industry because there are many food-grade lipids and emulsifiers available. During the past decade, many efforts have been devoted to the design and development of different lipid-based delivery systems, and significant progress has been made. The design principles are now quite clear, but further in depth studies are required on the in vivo biological efficacies and bioavailability of encapsulated active compounds. The potential use of these lipid-based nanoencapsulation systems in food products and their effect on human health has yet to be explored.

Chapter 4
Liquid-Based Nanoencapsulation Techniques

Nanoencapsulation is one of the most promising new technologies, having the feasibility to entrap bioactive compounds. It has versatile advantages in terms of targeted site-specific delivery and efficient absorption through cells. This chapter focuses on the various liquid-based nanoencapsulation techniques such as coacervation, inclusion encapsulation, nanoprecipitation, emulsification-solvent evaporation, and supercritical fluid. The current state of knowledge, limitations of these techniques, and recent trends are discussed.

4.1 Nanoprecipitation Technique

The different liquid-based techniques for nanoencapsulation of bioactive compounds are described in Table 4.1 (Ezhilarasi et al. 2013). The nanoprecipitation technique involves the precipitation of polymer from an organic solution and the diffusion of the organic solvent in an aqueous medium (Galindo-Rodriguez et al. 2004). Nanoprecipitation is based on the spontaneous emulsification of the organic internal phase containing the dissolved polymer, drug, and organic solvent into the aqueous external phase. The solvent displacement forms both nanocapsules and nanospheres. The biodegradable polymers commonly used are poly(ε-caprolactone) (PCL), poly(lactide) (PLA), poly(lactide-*co*-glycolide) (PLGA), Eudragit, and poly(alkylcyanoacrylate) (PACA) (Reis et al. 2006). Using nanoprecipitation along with freeze-drying techniques, Yallapu et al. (2010) encapsulated curcumin in PLGA in the presence of poly(vinyl alcohol) (PVA) and poly(L-lysine) stabilizers. The obtained particles were in the size range of 76–560 nm and encapsulation efficiency ranged from 49 % to 89 %. Curcumin nanoparticles showed an improved anticancer potential in cell proliferation and clonogenic assays, compared with free curcumin. Similarly, Anand et al. (2010) encapsulated curcumin in PLGA with polyethylene glycol (PEG)-5000 as stabilizer using the nanoprecipitation technique.

C. Anandharamakrishnan, *Techniques for Nanoencapsulation of Food Ingredients*,
SpringerBriefs in Food, Health, and Nutrition, DOI 10.1007/978-1-4614-9387-7_4,
© C. Anandharamakrishnan 2014

Table 4.1 Liquid-based nanoencapsulation techniques for bioactive compounds (Ezhilarasi et al. 2013)

Nanoencapsulation technique	Important raw materials used	Bioactive compound	Particle size	Purposes	References
Coacervation	Gelatin, maltodextrin, and tannins	Capsaicin	100 nm	Mask the pungent odor	Wang et al. (2008a)
	Gelatin, acacia, and hydrolysable tannins	Capsaicin	300–600 nm	Improve the efficiency and delay release	Xing et al. (2004)
	Gelatin, acacia, and tannins	Capsaicin	100 nm	Mask the pungent odor and improve the stability	Jincheng et al. (2010)
	Chitosan, poly(ethylene glycol-ran-propylene glycol)	BSA	200–580 nm	Control the release of encapsulated protein	Gan and Wang (2007)
Inclusion complexation	β-Lactoglobulin and low methoxyl pectin	DHA	100 nm	Improve stability and protection against degradation	Zimet and Livney (2009)
	α- and β-cyclodextrin	Linoleic acid	236 nm	Improve thermal stability	Hadaruga et al. (2006)
Nanoprecipitation	Monomethoxy poly(ethylene glycol)-poly(ε-caprolactone) micelles	Curcumin	27 nm	Improve solubility	Gou et al. (2011)
	PLGA	Curcumin	81 nm	Improve bioavailability and bioactivity, and enhance the cellular uptake	Anand et al. (2010)
	Ethyl cellulose and methyl cellulose	Curcumin	117 and 218 nm	Improve oral bioavailability and sustainability	Suwannateep et al. (2011)
	Poly(D,L-lactic acid) and PLGA	β-Carotene	80 nm	Improve physical and chemical stability and bioavailability	Ribeiro et al. (2008)
	Poly(ethylene oxide)-4-methoxyc innamoylphthaloylchitosan, poly(vinylalcohol-co-vinyl-4-ethoxycinnamate), PVA, and ethyl cellulose	Astaxanthin	300–320 nm	Improve the solubility and bioavailability	Tachaprutinun et al. (2009)
	PLGA	Curcumin	76–560 nm	Improve anticancer activity	Yallapu et al. (2010)
	Chitosan/poly(ε-caprolactone)	Curcumin	220 and 360 nm	Sustained release and increase the cellular uptake	Liu et al. (2012a)

Technique	Carrier	Active compound	Size	Objective	Reference
Emulsification-solvent evaporation	Chitosan crosslinked with tripolyphosphate	Curcumin	254–415 nm	Controlled release	Sowasod et al. (2008)
	Hydroxypropyl methyl cellulose and polyvinyl pyrrolidone	Curcumin	100 nm	Enhance absorption and prolong the rapid clearance of curcumin	Dandekar et al. (2010)
	Poly(D,L-lactide) and PVA	Quercetin	170 nm	Improve the controlled release and encapsulation efficiency	Kumari et al. (2010)
	Poly(methyl methacrylate) and PVA	Coenzyme Q_{10}	40–260 nm	Improve the reproducibility, stability, and target drug loading yield	Kwon et al. (2002)
	Tween 20	Phytosterol	50–282 nm	Optimize the operating conditions and reduce phytosterol loss	Leong et al. (2011)
	Tween 20	α-Tocopherol	90–120 nm	Minimize the recoalescence; improve the physical stability and solubility	Cheong et al. (2008)
	Sodium caseinate	Astaxanthin	115–163 nm	Optimize the processing conditions and improve bioavailability	Anarjan et al. (2011)
	Tween 20	β-Carotene	9–280 nm	Improve the physical stability	Silva et al. (2011)
	PLGA and PVA	Curcumin	45 nm	High yield, entrapment efficiency and sustained delivery	Mukerjee and Vishwanatha (2009)
	PLGA	Curcumin	158 nm	Improve oral bioavailability	Tsai et al. (2011)
	PLGA	Curcumin	200 nm	Increase water solubility and bioavailability	Xie et al. (2011)
	PLGA and PEG	Curcumin	200 nm	Increase bioavailability	Khalil et al. (2013)
Supercritical antisolvent precipitation	Hydroxylpropyl methyl cellulose phthalate	Lutein	163–219 nm	Higher lutein loading and encapsulation efficiency	Jin et al. (2009)

BSA bovine serum albumin, *PLGA* poly(lactide-*co*-glycolide), *PEG* polyethylene glycol, *PVA* polyvinyl alcohol

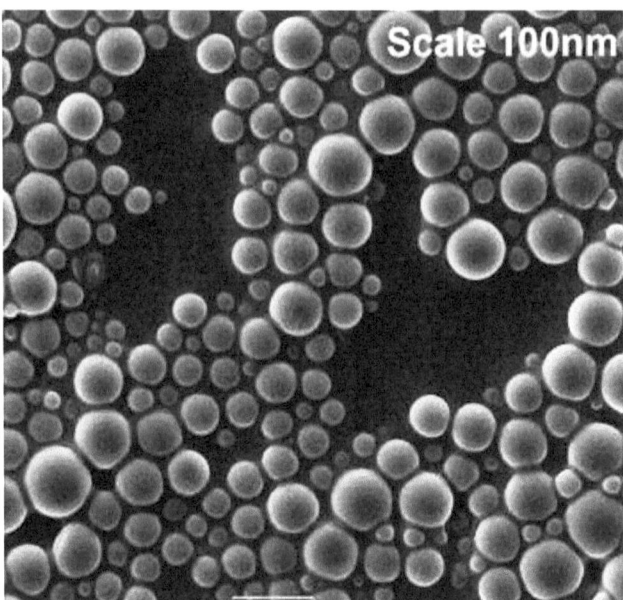

Fig. 4.1 Scanning electron micrograph of nanoencapsulated curcumin (Anand et al. 2010)

The nanocapsules had a mean particle diameter of about 81 nm (Fig. 4.1) and were reported to have enhanced cellular uptake. Moreover, the curcumin nanocapsules exhibited increased bioactivity (in vitro) and superior bioavailability (in vivo) compared with curcumin. Recently, Gou et al. (2011) encapsulated curcumin using a single-step nanoprecipitation method along with freeze-drying. The nanocapsules had a mean particle size of 27 nm, with 99 % encapsulation efficiency. They also exhibited a stronger anticancer effect than free curcumin in in vivo studies. Using a simple nanoprecipitation method, curcumin-loaded chitosan/poly(ε-caprolactone) (chitosan/PCL) spherical nanoparticles (220–360 nm) were obtained (Liu et al. 2012a). In vitro release studies showed the sustained release behavior of curcumin from nanoparticles during the 5 days study. Furthermore, in vitro cell uptake studies revealed that the cell uptake of curcumin was greatly enhanced by encapsulating curcumin into cationic chitosan/PCL. The encapsulation efficiency and loading capacity of nanoparticles were 71 % and 4 %, respectively. Likewise, Suwannateep et al. (2011) encapsulated curcumin in ethyl cellulose and a dipolymeric carrier (ethyl cellulose and methyl cellulose) using a solvent displacement method in combination with freeze-drying. The obtained curcumin-loaded ethyl cellulose(C-EC) and dipolymeric carriers had mean diameters of 281 and 117 nm, respectively. The curcumin nanocapsules showed in vitro cytotoxicity towards cancer cell lines, and C-EC exhibited excellent mucoadhesive properties.

Using the solvent displacement method (a schematic flow diagram is shown in Fig. 4.2), Ribeiro et al. (2008) produced β-carotene-loaded nanodispersions by

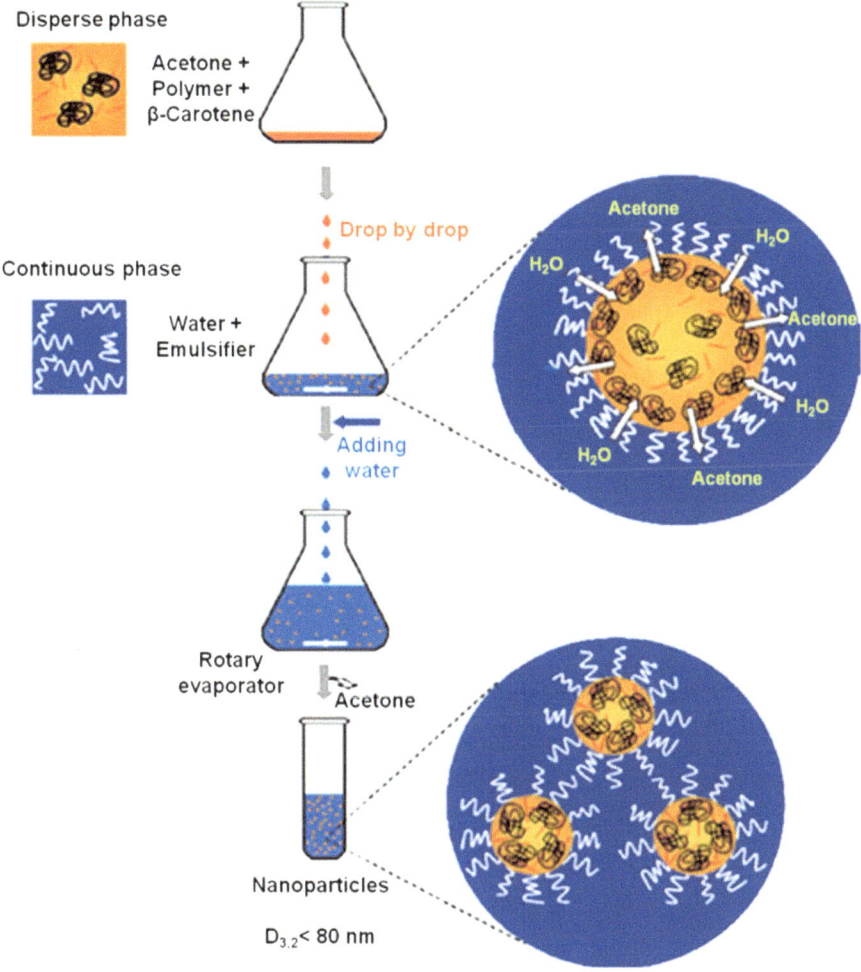

Fig. 4.2 Production of β-carotene nanodispersion by the solvent displacement method (Ribeiro et al. 2008)

encapsulating β-carotene into PLA and PLGA and subsequent freeze-drying. Gelatin and Tween 20 were used as stabilizing hydrocolloids in the continuous phase. The nanodispersion had a mean droplet diameter of below 80 nm with narrow size distribution. It also exhibited higher stability against Ostwald ripening and coalescence. Moreover, on redispersion of lyophilized powder in water, there was no significant change in droplet size. Tachaprutinun et al. (2009) encapsulated astaxanthin by solvent displacement along with freeze-drying. The method yielded reasonably good encapsulation efficiency (98 %) at a loading of 40 %. Moreover, the freeze-dried astaxanthin-loaded nanospheres showed good dispersibility in water and yielded stable aqueous suspensions of nanoparticles of about 300–320 nm. The nanospheres exhibited steady release of astaxanthin up to a maximum of 85 % payload over 60 min. NMR analysis after a 2-h heating at 70 °C showed minimal heat degradation of olefinic functionality in nanoencapsulated astaxanthin compared with unencapsulated pigment molecules, which were almost completely destroyed. Recently, Khayata et al. (2012) prepared vitamin-E-loaded nanocapsules (165–172 nm) by a nanoprecipitation method using the membrane contactor technique, with high encapsulation efficiency of about 98 %.

The nanocapsules exhibited good stability against degradation, higher encapsulation efficiency, sustained release, and increased cellular uptake and bioavailability during in vivo studies. However, the results depend on a good drying technique (freeze drying) and only polymer-based wall material can be used (PEG and PLGA). Appropriate solvent and non solvent phases need to be selected, which may vary for each bioactive component. The usefulness of this simple technique is limited to water-miscible solvents, in which the diffusion rate is enough to produce spontaneous emulsification. Thus, the nanoprecipitation technique can be used for production of nanocapsules of around 100 nm and below.

4.2   Emulsification-Solvent Evaporation Technique

The emulsification-solvent evaporation technique involves emulsification of the polymer solution into an aqueous phase and evaporation of polymer solvent, inducing polymer precipitation as nanospheres (Reis et al. 2006). The size of the capsules is influenced by the stir rate, type and amount of dispersing agent, viscosity of organic and aqueous phases, and temperature (Tice and Gilley 1985). Frequently used polymers are PLA, PLGA, ethyl cellulose (EC), cellulose acetate phthalate, PCL, and β-hydroxybutyrate (Ezhilarasi et al. 2013). In order to produce small particle size, high-speed homogenization or ultrasonication are often employed (Zambaux et al. 1998). Using a multiple emulsion/solvent technique along with freeze-drying, Sowasod et al. (2008) encapsulated curcumin in chitosan by crosslinking with tripolyphosphate. The curcumin nanocapsules were spherical in shape with particle sizes ranging from 254 to 415 nm. The yield of nanoencapsulated curcumin ranged from 19 % to 96 % and the Fourier transform infrared spectroscopy (FTIR) analysis confirmed the crosslinking between tripolyphosphate and the amine group of chitosan in nanocapsules. Likewise, Dandekar et al. (2009) encapsulated curcumin in Eudragit S 100 (polymer) using a solvent evaporation method followed by freeze-drying. The obtained nanocapsules were spherical, with an encapsulation efficiency of about 72 %. Moreover, the nanocapsules exhibited almost double the inhibition of cancerous cells compared with curcumin alone. Similarly, using the emulsification-solvent evaporation technique, Mukerjee and Vishwanatha (2009) prepared curcumin-loaded PLGA nanospheres. The nanospheres were smooth and spherical, with particle size distribution in the range of 35–100 nm, and mean particle diameter of 45 nm. The nanocapsules exhibited high yield and drug

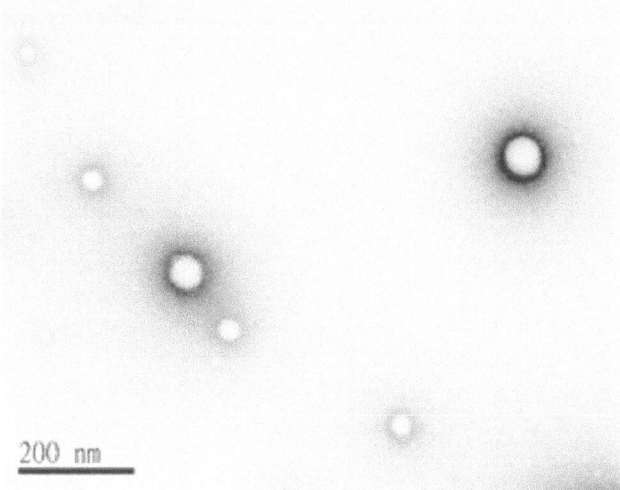

Fig. 4.3 Transmission electron microscopy of curcumin-loaded PLGA nanoparticles (Tsai et al. 2011)

entrapment efficiency as well as sustained delivery of curcumin. Further, the authors reported higher intracellular uptake and efficient action in prostate cancer cell lines. Recently, Dandekar et al. (2010) also encapsulated curcumin in a nanocarrier using the solvent emulsion-evaporation technique. These nanoparticles were observed to be of around 100 nm in size, with a fairly narrow distribution and encapsulation efficiency of 72 %. This optimized system was further subjected to freeze-drying. The freeze-dried product, on reconstitution, exhibited a size and distribution similar to that before freeze-drying. In vivo anti-malarial studies revealed significantly superior action of nanoparticles compared with curcumin control.

Tsai et al. (2011) obtained curcumin-loaded PLGA nanoparticles with a mean particle size of 158 nm (Fig. 4.3) using high pressure emulsification-solvent evaporation and subsequent freeze-drying. Oral bioavailability of curcumin-loaded nanocapsules was 22-fold higher than that of conventional curcumin. Likewise, Xie et al. (2011) encapsulated curcumin with PLGA. The obtained nanoparticles exhibited a smooth and spherical shape with mean diameters of about 200 nm. The entrapment efficiency and loading rate were 92 % and 6 %, respectively. Nanoparticles showed about 640-fold higher water solubility relative to that of native curcumin. Recently, Khalil et al. (2013) produced PLGA and PLGA–PEG blend nanoparticles (<200 nm) containing curcumin by a single-emulsion solvent evaporation technique with sonication and subsequent freeze-drying. In vitro release studies showed that curcumin was released more slowly from the PLGA nanoparticles than from the PLGA–PEG nanoparticles. In the pharmacokinetic study, compared to curcumin aqueous suspension, the PLGA and PLGA–PEG nanoparticles increased the curcumin bioavailability by 16- and 55-fold, respectively.

Fig. 4.4 Preparation of
nanoparticles by
emulsification-solvent
evaporation method
(Kwon et al. 2002)

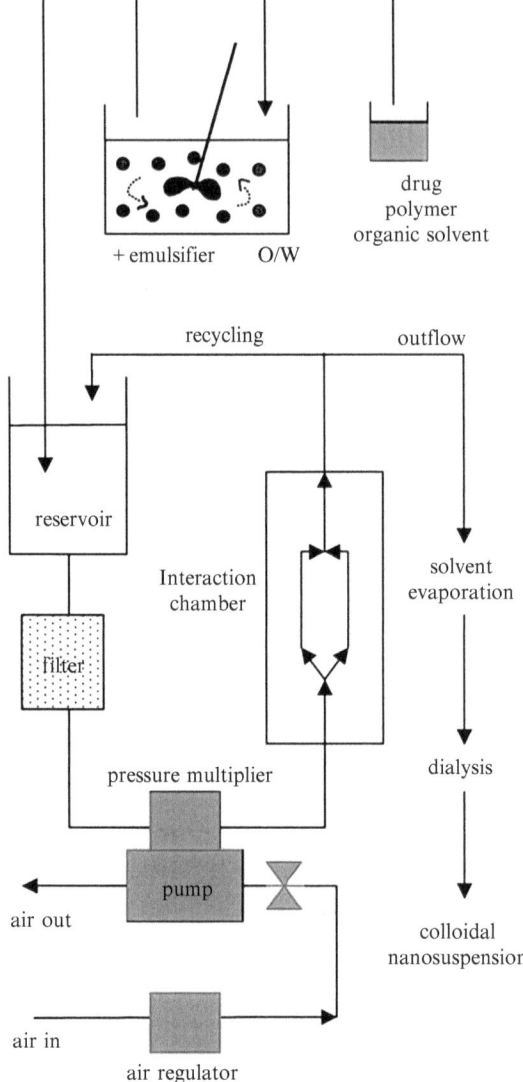

Using emulsification-solvent evaporation, Kwon et al. (2002) prepared coenzyme Q_{10}-loaded nanoparticles (40–260 nm) (Fig. 4.4). Despite, a very high target
drug loading yield of around 39 %, the actual loading efficiency reached above
95 %, and the mean diameter of the nanoparticles was highly influenced by the
kind of surfactants used and the recycling number of the microfluidization process. Recently, Kumari et al. (2010) encapsulated quercetin in PLA by solvent
evaporation followed by freeze-drying. The nanoparticles had a mean diameter of

about 130 nm with encapsulation efficiency of 97 %. The in vitro release kinetics under physiological conditions exhibited an initial burst release followed by slow and sustained release. The complete release and maximum release of 88 % quercetin were at 72 and 96 h, respectively. Continuing similar work, Kumari et al. (2011) encapsulated quercitrin in PLA by a solvent evaporation method to improve the solubility, permeability, and stability of this molecule. The nanocapsules had mean particle size of about 250 nm and encapsulation efficiency of about 40 %. Their in vitro release kinetics under physiological condition revealed an initial burst release followed by sustained release of quercitrin. Less fluorescence quenching was also observed with equimolar concentration of PLA-encapsulated quercitrin than with free quercitrin. The presence of quercitrin-specific peaks in the FTIR spectra of washed (five times) quercitrin-loaded PLA nanoparticles provides extra evidence for the encapsulation of quercitrin into PLA nanoparticles. Cheong et al. (2008) prepared a α-tocopherol nanodispersion using emulsification evaporation under various combinations of the processing parameters and the ratio of aqueous to organic phases. The nanodispersions had mean droplet diameters in the range of 90–120 nm and there were no significant changes in mean diameters during a storage period of 3 months. Moreover, the processing cycle did not exhibit a significant effect on the droplet diameter and size distribution of the prepared nanodispersion.

Similarly, using emulsification evaporation, Anarjan et al. (2011) prepared a nanodispersion (110–165 nm) of astaxanthin. The most desirable nanodispersion was obtained using a high-pressure homogenization at 30 MPa with three passes, followed by evaporation at 25 °C. Leong et al. (2011) obtained phytosterol nanodispersions of about 50–282 nm through emulsification evaporation using high-pressure homogenization. Phytosterol loss after high-pressure homogenization ranged from 3 % to 28 %, and losses increased with increasing homogenization pressure. Recently, Silva et al. (2011) produced nanoemulsions of β-carotene (9–280 nm) using a high-energy emulsification-evaporation technique. The process parameters such as time and shear rate of homogenization significantly affected particle size distribution and storage stability of nanoemulsions. β-Carotene nanoemulsions showed good physical stability in terms of size distribution but were chemically unstable during storage (evaluated in terms of β-carotene retention), which was observed through the color of the nanoemulsions. Emulsification-solvent evaporation was observed to be an efficient technique for producing nanocapsules of below 100 nm. Most of the nanocapsules exhibited a spherical shape, high drug loading content, and encapsulation efficiency of about 75–96 % with sustained release and increased absorption. Apart from this, the nanodispersions and nanoemulsions prepared through this method exhibited good stability. However, the method depends on a suitable emulsification technique such as microfluidization, and high-speed and high-pressure homogenization techniques. It also relies on a suitable drying technique for producing nanocapsules (Ezhilarasi et al. 2013).

4.3 Inclusion Complexation

Inclusion complexation is defined as encapsulation of the supramolecular association of a ligand (encapsulated ingredient) into a cavity-bearing substrate (shell material) through hydrogen bonding, van der Waals force, or by the entropy-driven hydrophobic effect (Ezhilarasi et al. 2013). In the food industry, molecular entities having suitable molecular-level cavities are rarely available. Thermal stability of linoleic acid was improved by encapsulating it in α- and β-cyclodextrin (α- and β-CD) using the inclusion complexation technique (Hadaruga et al. 2006). A relative concentration above 98 % fatty acid was reported in the case of temperature degradations of 50 °C and 100 °C, but at 150 °C only 92 % was observed and linoleic acid was partially converted to more stable geometrical isomers. The nanocapsules of α- and β-CD complexes had yields of about 88 % and 74 %, respectively. Similarly, usnic acid (UA) was encapsulated in β-CD by inclusion complexation along with freeze-drying (Lira et al. 2009). The complex (UA:β-CD) was further incorporated into liposomes using hydration of a thin lipid film with subsequent sonication in order to produce a targeted drug delivery system. Liposomes containing UA:β-CD exhibited drug encapsulation efficiency (99.5 %) and remained stable for 4 months in a suspension form. Liposomes containing UA:β-CD presented a more prolonged release profile of free usnic acid than usnic acid-loaded liposomes (Lira et al. 2009). Another example of molecular inclusion used the milk protein β-lactogloglobulin (β-Lg). Zimet and Livney (2009) produced a stable (colloidal) nanocomplex of docosahexaenoic acid (DHA)-loaded β-Lg along with low methoxy pectin. The nanocomplex of DHA-loaded β-Lg provided good protection against degradation of DHA during an accelerated shelf-life stress test. It was reported that only about 5–10 % was lost in the nanocomplex during 100 h at 40 °C compared to about 80 % lost in the unprotected DHA. The inclusion complexation technique is mainly used in the encapsulation of volatile organic molecules (essential oils and vitamins). It is useful for the masking of odors and flavors and to preserve aromas. This technique yielded higher encapsulation efficiency with higher stability of the core component. However, only a few particular molecular compounds like β-cyclodextrin and β-lactogloglobulins are suitable for encapsulation through this method.

4.4 Coacervation

Coacervation is a distinctive and promising encapsulation technology because very high payloads are achievable (up to 99 %) and it allows the possibility of controlled release based on mechanical stress, temperature, or sustained release (Gouin 2004). The coacervation technique involves the phase separation of a single polyelectrolyte or mixture of polyelectrolytes from a solution and subsequent deposition of the newly formed coacervate phase around the active ingredient. Further, a hydrocolloid shell can be crosslinked using an appropriate chemical or enzymatic crosslinker

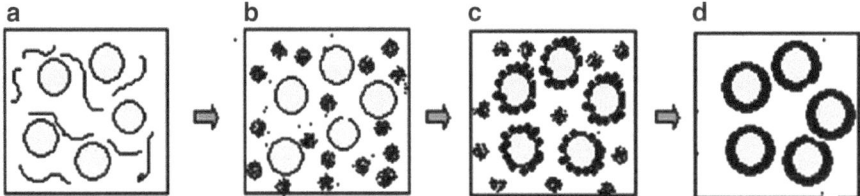

Fig. 4.5 Formation of the nanoencapsulated capsaicin agents: (**a**) dispersion of capsaicin in gelatin solution; (**b**) coacervation of gelatin with acacia in the solution; (**c**) coacervation of insoluble complex on the surface of the capsaicin; and (**d**) shell formation by the addition of glutaraldehyde solution (Jincheng et al. 2010)

such as glutaraldehyde or transglutaminase, mainly to increase the strength of the coacervate (Zuidam and Shimoni 2010). Based on the number of polymer types used, the process can be termed a simple coacervation (only one type of polymer) or a complex coacervation (two or more types of polymer). Many factors including the biopolymer type (molar mass, flexibility, and charge), pH, ionic strength, concentration, and ratio of the biopolymers affect the power of the interaction between the biopolymers and nature of the complex formed (Tolstoguzov 2003; De Kruif et al. 2004; Turgeon et al. 2007)

Apart from the electrostatic interactions between biopolymers of opposite charge, hydrophobic interactions and hydrogen bonding can also contribute significantly to complex formation. Xing et al. (2004) encapsulated capsaicin in gelatin and acacia using a complex coacervation technique. The nanocapsules were obtained by treating encapsulated capsaicin with hydrolysable tannins, crosslinking with glutaraldehyde, and subsequent freeze-drying. The mean diameter of nanocapsules was found to be 300–600 nm with a spherical morphology. The nanocapsules exhibited 81 % encapsulation efficiency, 21 % drug loading capacity, and good dispersion.

Simple coacervation was used for encapsulation of capsaicin in gelatin by crosslinking with glutaraldehyde and followed by drying in a vacuum oven (Wang et al. 2008a). The obtained nanocapsules were of 100 nm size. The optimized conditions for the synthesis of nanocapsules could be related to the process conditions, such as higher shearing force (16,000 rpm agitation rate), lower gelatin viscosity (10–15 cP polymer viscosity), suitable crosslinking time (30–60 min), the utilization of tannins, and other experimental conditions beneficial for the fabrication of a thinner capsule wall. Moreover, the melting point and thermal pyrolysis temperature of the nanoencapsulates were improved due to encapsulation of the crosslinked gelatin over the surface of the capsaicin. Jincheng et al. (2010) encapsulated capsaicin by the complex coacervation method, as shown in Fig. 4.5. The authors observed that higher shearing force (15,000 rpm agitation rate), lower gelatin viscosity (15–20 cP), suitable crosslinking time (40–80 min), utilization of tannins, and the other experimental conditions influenced the synthesis of nanoencapsulates. The obtained nanoencapsulates had a mean diameter of about 100 nm with a spherical morphology, increase in melting point (75–85 °C), and improvement in degradation properties.

Their results were consistent with the earlier studies. Gan and Wang (2007) encapsulated a model protein, bovine serum albumin (BSA), in chitosan by an incorporation or incubation method using the polyanion tripolyphosphate (TPP) as the coacervation crosslinking agent. The BSA-loaded chitosan–TPP nanoparticles prepared under varying conditions were found to be in the size range of 200–580 nm.

The coacervation technique produced nanocapsules in the size range of 100–600 nm. However, the method depends on a suitable drying technique such as vacuum drying or freeze-drying. Gelatin, gum acacia, and chitosan are mostly used as a wall material in this technique. Moreover, the morphology (good dispersion and shape) and particle distribution of the nanocapsules was influenced by treatment with tannins. The thermal stability and melting point of the nanocapsules was increased by crosslinking with glutaraldehyde (Ezhilarasi et al. 2013).

4.5 Supercritical Fluid Technique

Supercritical fluids exhibit properties intermediate between those of liquids and gases, such as low viscosity, low density, high solvating power, high diffusivities, and high mass transfer rates above the critical point. A number of compounds can be brought to a supercritical state, such as carbon dioxide, water, propane, nitrogen, etc. (Gouin 2004). Some of the methods using supercritical fluid technology are rapid expansion from supercritical solution, gas anti-solvent, supercritical anti-solvent precipitation, aerosol solvent extraction, and precipitation with a compressed fluid anti-solvent (Kikic et al. 1997). Supercritical fluids are used for encapsulation of thermally sensitive compounds in a process similar to spray drying. In this technique, the bioactive compound and the polymer are solubilized in a supercritical fluid and the solution is expanded through a nozzle. The supercritical fluid is evaporated in the spraying process, and solute particles eventually precipitate (Reis et al. 2006). This technique has been widely used for low critical temperatures and minimum use of organic solvent (Ezhilarasi et al. 2013).

Jin et al. (2009) encapsulated lutein in hydroxypropyl methyl cellulose phthalate (HPMCP), using supercritical anti-solvent precipitation (CO_2 as supercritical fluid), to maintain its bioactivity and avoid thermal and light degradation. Various operating parameters such as lutein loading, encapsulation efficiency, and particle size affected product yield. The obtained lutein-loaded HPMCP nanocapsules had mean diameter in the range of 163–219 nm (Fig. 4.6). The highest lutein-loading (16%) and encapsulation efficiency (88 %) were obtained under the operating conditions of 11 MPa pressure at 40 °C temperature with a 5:1 ratio of HPMCP and lutein. Turk and Lietzow (2004) synthesized phytosterol nanoparticles (size below 500 nm) using rapid expansion from supercritical solution (CO_2 as supercritical fluid). The surfactant type and concentration were reported to influence the particle size distribution. However, supercritical fluid technology requires high initial capital investment for high-pressure equipment (Gouin 2004). Besides these techniques, electrospinning, electrospraying, and emulsification are also used for nanoencapsulation, which are discussed in separate chapters.

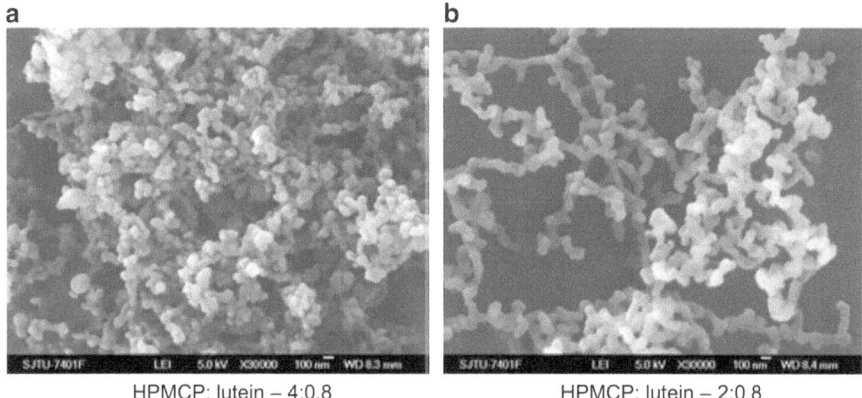

HPMCP: lutein – 4:0.8   HPMCP: lutein – 2:0.8

Fig. 4.6 Scanning electron micrograph of nanocapsules prepared using hydroxyl propyl methyl cellulose phthalate and lutein at two different ratios (Jin et al. 2009)

Chapter 5
Electrospraying and Electrospinning Techniques for Nanoencapsulation

In recent years, electrospraying and electrospinning have attracted widespread interest and found applications in the food industry and for drug delivery. These electrohydrodynamic processes are having great potential for making micro- and nanosized particles and fibers. However, more research is needed to optimize the operating conditions for the nanoencapsulation of various food and bioactive compounds. In addition, the properties, and efficiency of the nanocapsules have to be analyzed to improve their application in the food industry. The current state of knowledge, limitations of electrospinning and electrospraying techniques, and recent trends are discussed.

Electrospraying and electrospinning techniques are very cost-effective and use a uniform electrohydrodynamic force to break the liquids into fine jets (Wu and Clark 2008). Electrospraying and electrospinning are promising techniques, attracting researchers' interest and curiosity due to their possible applications in food materials and drug delivery. Both techniques work on the same principle with very minor and basic differences. In electrospinning, the polymer is transformed into continuous nanosized polymer threads, whereas electrospraying results in nanosized particles of the polymer. The latter is modified version of the former (Arya et al. 2008). The application of electrospinning and electrospraying techniques for nanoencapsulation of bioactive compounds is shown in Table 5.1.

5.1 Electrospraying

The polymer solution (with low viscosity) is allowed to pass through a syringe under the influence of an electric field to generate a fine mist. The applied voltage forces the solution to come out of the syringe in the form of a jet and the formation of micro-/nanoparticles takes place. The liquid, after acquiring charge at the end of the nozzle, forms a Taylor cone. Liquid is forced to endure more and more charge.

C. Anandharamakrishnan, *Techniques for Nanoencapsulation of Food Ingredients*,
SpringerBriefs in Food, Health, and Nutrition, DOI 10.1007/978-1-4614-9387-7_5,
© C. Anandharamakrishnan 2014

Table 5.1 Electrospraying and electrospinning techniques for nanoencapsulation of bioactive compounds

Technique used	Wall material used	Core material	Size	Purpose	Reference
Electrospraying	Zein ultrathin fibers	DHA	500–700 nm	Improved stability	Torres-Giner et al. (2010)
	Chitosan micro-/nanospheres	Ampicillin sodium	455–885 nm	Higher encapsulation efficiency and improved anti-bacterial activity	Arya et al. (2008)
	Whey protein concentrate	Carotene	<100 nm	Whey protein explored as the coat material to deliver bioactives	Lopez-Rubio and Lagaron (2012)
	Zein	Curcumin	175–250 nm	Enhanced stability and dispersion in aqueous food matrix	Gomez-Estaca et al. (2012)
Electrospinning	Ultrathin PVA	Bifidobacteria	150 nm	Improved viability of bacteria	Lopez-Rubio et al. (2009)
	PVA	Curcumin	250–350 nm	Enhanced stability of curcumin	Sun et al. (2013)
	Tecophilic polyurethane	BSA and epidermal growth factor	200 nm	Encapsulation of two compounds as separated domains within a single fiber	Dong et al. (2009)
	Zein prolamine	Beta-carotene	1,140 nm	Enhanced oxidative stability and light stability	Fernandez et al. (2009)
	Zein	Curcumin	310 nm	Improved sustained release and effective free radical scavenging ability	Brahatheeswaran et al. (2012)
	Cellulose acetate	Curcumin	314–340 nm	Stable chemical integrity and the product was confirmed to be nontoxic to human dermal fibroblasts	Suwantong et al. (2007, 2010)
	PVA and cyclodextrin	Vanillin	120–230 nm	Efficient thermal stability	Kayaci and Uyar (2012)

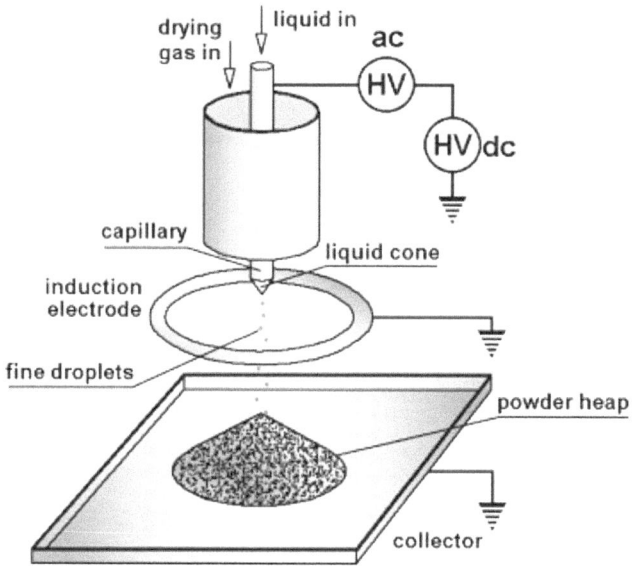

Fig. 5.1 Schematic illustration of electrospraying (Jaworek 2007)

After a certain limit, the liquid disperses into numerous micron-sized droplets and falls onto the surface with opposite charge. As these droplets move, the solvent at their surface evaporates and shrinkage occurs (Chen 2007) (as shown in Fig. 5.1).

A relationship was observed between the electrical pressure resulting from excess charge q on a spherical droplet of radius r and surface tension σ. It was predicted that the natural quadrupolar oscillation of a droplet in a field-free environment becomes unstable when q exceeds the limit q_R, known as the "Rayleigh limit" (Grimm and Beauchamp 2005):

$$q_R = 8\pi\varepsilon^{1/2}\sigma^{1/2}r^{3/2} \tag{5.1}$$

The limit is reached either by evaporation or by application of charge in excess of q_R. The size of the droplets formed purely depends on the flow rate of liquid and the charge applied by varying the voltage. This technique can be used for producing monodisperse particles.

Arya et al. (2008) optimized the various factors of the electrospraying technique for producing ampicillin-loaded chitosan micro-/nanospheres. The major factors considered were electrospraying voltage, needle gauge, concentration, and electrospraying distance. It was reported that 7 cm working distance and 26 g of needle gauge produced nanoparticles with 128.2 mV zeta potential and 80 % encapsulation efficiency, with improved stability. Similarly, Torres-Giner et al. (2010) encapsulated an omega-3 fatty acid (docosahexaenoic acid, DHA) by electrospraying using ultrathin zein. It was reported that the nanocapsules were more stable across changes in relative humidity and temperature.

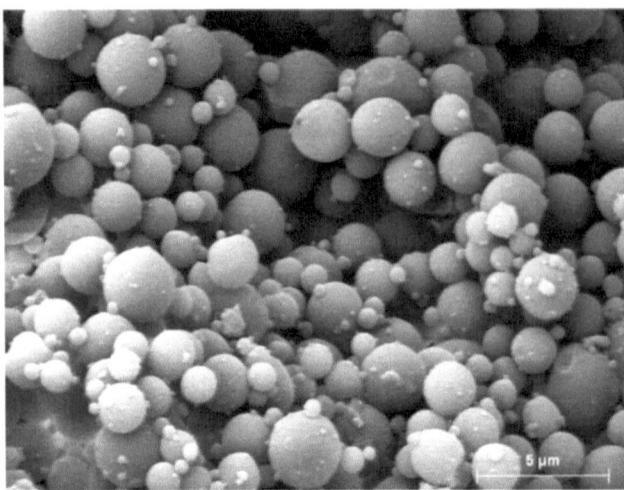

Fig. 5.2 Scanning electron micrographs of whey protein concentrate (WPC) capsules obtained through electrospraying from 40 wt% WPC aqueous solutions (Lopez-Rubio and Lagaron 2012)

Recently, Lopez-Rubio and Lagaron (2012) encapsulated β-carotene using whey protein concentrate (WPC) at the micro-, submicro- and nanoscales by electrospraying. The capsules had a mean diameter of less than 100 nm (as shown in Fig. 5.2). Moreover, the concentration of whey proteins, range of pH, and additional component glycerol influenced the size and morphology of capsules. Similarly, Gomez-Estaca et al. (2012) electrosprayed zein solutions containing curcumin and obtained spherical nanoparticles. The nanoparticles possessed good encapsulation efficiency (85–90 %) and their diameters ranged between 175 and 250 nm. Further, electrospraying offered enhanced stability and dispersibility to the active compound.

5.2   Electrospinning

Electrospinning is a cheap and unique method of producing polymer fibers in the range of 100 nm diameter. It works on the same principle as electrospraying. But, in electrospinning high voltage is applied, which solidifies the liquid and generates the polymer fiber (Chen 2007). A schematic set-up for electrospinning is shown in Fig. 5.3.

In this method, two electrodes are used: one is connected to the spinning solution and other is attached to a collector, which is a metallic foil. A droplet of the polymer solution is suspended in the tip of the needle. The electric charge overcomes the surface tension of the droplet and a charged jet is emitted. The jet passes the set distance (approximately 15 cm) between the needle tip and the collector. Between the needle and collector, the nanofibers are submitted to a combined stretch effect,

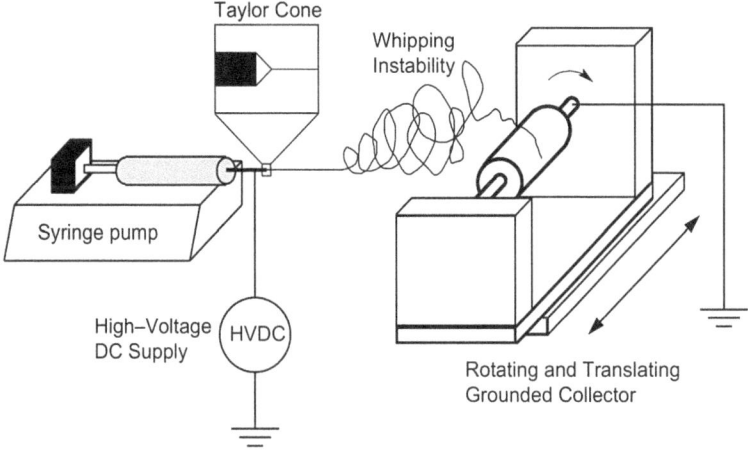

Fig. 5.3 Schematic illustration of electrospinning (Sill and von Recum 2008)

Table 5.2 Effects of electrospinning parameters on fiber morphology (Sill and von Recum 2008)

Parameter	Effect on fiber morphology
Applied voltage ↑	Fiber diameter ↓ initially, then ↑ (monotonic)
Flow rate ↑	Fiber diameter ↑ (beaded morphologies occur if the flow rate is too high)
Distance between capillary and collector ↑	Fiber diameter ↓ (beaded morphologies occur if the distance between the capillary and collector is too short)
Polymer concentration (viscosity) ↑	Fiber diameter ↑ (with optimal range)
Solution conductivity ↑	Fiber diameter ↓ (broad diameter distribution)
Solvent volatility ↑	Fibers exhibit microtexture (pores on their surfaces, which increase the surface area)

firstly a radial stretch caused by repulsion between the positively charged polymer chains and, secondly, a longitudinal stretch caused by the high collector attraction (Chen 2007).

The factors that influence the process are voltage, solution flow rate, polymer weight concentration, molecular weight of the polymer, and nozzle-to-ground distance. Based on these parameters, Sill and von Recum (2008) (Table 5.2) and Chen (2007) made the following observations:

- At constant electric potential, flow rate, and concentration, the fiber diameter is inversely proportional to screen distance (cm)
- At constant flow rate, screen distance, and concentration, the fiber diameter is inversely proportional to electric potential (kV)
- At constant electric potential and screen distance, the fiber diameter is directly proportional to flow rate (mL/h) when concentration is constant and to concentration (wt%) when flow rate is constant

Apart from process parameters, certain solution characteristics also influence the fiber formation and structure. These parameters include polymer concentration, solvent volatility, and solvent conductivity (Sill and von Recum 2008):

- Polymer concentration is directly proportional to viscosity and surface tension. For better results, a solution must have high polymer concentration.
- Choice of solvent is another major factor that influences fiber formation. Volatility of the solvent should be adequate to allow sufficient solvent evaporation between the tip and collector. The solvent volatility also influences the phase separation during the flight of the solvent.
- Solution conductivity was observed to have less influence on the process and within 1–2 orders of magnitude. It directly affects the tensile strength of the fiber produced. The fiber jet of higher conductivity solutions possesses higher tensile strength.

Electrospinning can be performed in two different ways, either through coaxial electrospinning or through direct incorporation of the material with the polymeric solution, with different applications. Electrospinning has spread its roots into the field of tissue engineering and has shown a high potential in the engineering of a number of tissues, including vasculature, bone, neural, and tendon/ligament.

Delivery of drugs or bioactive compounds is another field of application for electrospinning. The bioactive compound can be encapsulated in the fiber for controlled release at the target site. Two-phase electrospinning has also been used for encapsulation. This procedure involves a biphasic suspension, formed by mixing an aqueous solution of the biological material with an organic-polymer solution. Then, the suspension is electrospun, resulting in the nanoencapsulation of aqueous reservoirs within the polymer fibers. Two-phase electrospinning has been used to encapsulate proteins, including growth factors and cytochrome C, into biocompatible polymers (Dong et al. 2009). The size of the encapsulated drug produced may be either micro- or nanosized. In the medical field, coaxial electrospinning has been applied for the incorporation of bovine serum albumin (BSA) into biodegradable poly(ε-caprolactone) (PCL) (Zhang et al. 2006).

Dong et al. (2009) exploited the application of electrospinning for incorporating two different model proteins, such as BSA fluorescently labeled with Texas Red and epidermal growth factor fluorescently labeled with AlexaFluor 488, into the same coat material in two different domains. Tecophilic polyurethane and poly(lactic-co-glycolic acid) were used as coat material. The procedure involved three steps: (i) direct synthesis of two sets of poly(vinyl alcohol) (PVA) nanoparticles, each containing a different biomolecule in a common solution of a biocompatible polymer; (ii) mixing of these nanoparticle-containing solutions together; and (iii) electrospinning of the combination. The average size of the nanoparticles was 200 nm. Similarly, Lopez-Rubio et al. (2009) encapsulated *bifidobacteria* (with average diameter of 150 nm) in ultrathin PVA electrofibers. The nanocapsules showed a

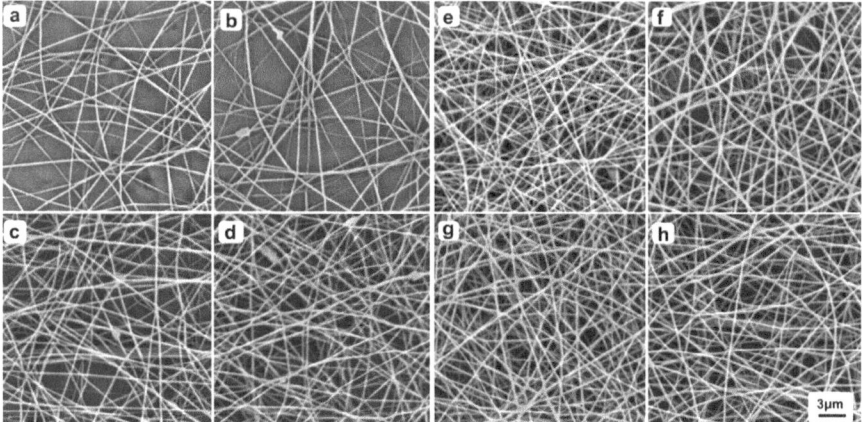

Fig. 5.4 Scanning electron micrographs of curcumin-loaded PVA nanofibers with (**a**) 5, (**b**) 10, (**c**) 15, and (**d**) 20 wt% drug content, and complex-loaded fibers containing (**e**) 20, (**f**) 30, (**g**) 40, and (**h**) 50 wt% complex (Sun et al. 2013)

high oxygen barrier and resulted in an increased shelf life of the bacteria. Recently, using electrospinning, Sun et al. (2013) loaded curcumin in PVA and β-cyclodextrin through inclusion complexation. Encapsulated curcumin nanofibers had an average diameter of 250–350 nm (as shown in Fig. 5.4). In this study, PVA/complex fibers showed faster release than PVA/curcumin. A light-sensitive compound such as β-carotene was encapsulated in ultrafine fibers of zein prolamine by Fernandez et al. (2009). The fibers obtained had mean diameters in the nanorange (1,140 nm) and the encapsulated compound was proved to possess increased light stability and oxidative stability.

Curcumin, a food bioactive ingredient, has been encapsulated into nanofibers by various researchers for drug delivery applications. For example, Brahatheeswaran et al. (2012) encapsulated curcumin in to zein nanofibers and obtained smooth-surfaced encapsulates with an average diameter of 310 nm. The compound was shown to have free radical scavenging activity and was effective in sustained release of the bioactive compound. In another study, the nanofibers obtained by electrospinning curcumin and cellulose acetate (approximately 314–340 nm) retained their chemical integrity for a storage period of 4 months and the product was confirmed to be nontoxic to human dermal fibroblasts (Suwantong et al. 2007, 2010). Apart from these, volatiles and aroma compounds have also been encapsulated successfully using electrospinning techniques. Kayaci and Uyar (2012) formed nanofibers of PVA containing a vanillin-cyclodextrin inclusion complex. The average diameters were in the range of 120–230 nm and the encapsulated vanillin was found to be thermally stable.

Chapter 6
Drying Techniques for Nanoencapsulation

Nanoencapsulation techniques are used to produce nanosuspensions of active compounds with a coating or encapsulated with wall materials, in liquid or dried form. The major problems of nanocapsules are irreversible aggregation and chemical instability and leakage of the encapsulated active ingredients. Therefore, it is desirable to convert nanocapsule suspensions into dried form to maintain their stability. The drying of nanoparticles facilitates easier handling and storage and they are readily dispersible in aqueous solutions. Hence, nanoencapsulation methods in combination with drying techniques are essential for converting encapsulated suspensions to a dried, stable form. This chapter reviews the various drying techniques for nanoencapsulation.

Freeze-drying and spray-drying techniques are commonly employed for drying of nanosuspensions and it is clear that the operating conditions of spray drying and freeze-drying are significantly important in the stabilization of nanocapsules. Besides higher stability compared to the original nanoparticle suspension, the dried powders have the ability to control and promote sustained bioactive compound release. However, drying aggravates additional stress on the nanocapsules during processing. This chapter discusses in detail the different drying techniques used for production of nanoparticles and described in table 6.1 (Ezhilarasi et al. 2013).

C. Anandharamakrishnan, *Techniques for Nanoencapsulation of Food Ingredients*,
SpringerBriefs in Food, Health, and Nutrition, DOI 10.1007/978-1-4614-9387-7_6,
© C. Anandharamakrishnan 2014

Table 6.1 Drying techniques for nanoencapsulation of bioactive compounds (Ezhilarasi et al. 2013)

Nanoencapsulation technique	Important raw materials used	Bioactive compound	Particle size	Purposes	References
Spray drying	Carbohydrate matrix and maltodextrin	Catechin	80 nm	Increase the stability and protect from oxidation	Ferreira et al. (2007)
	Modified n-octenyl succinate-starch	β-Carotene	300–600 nm (droplet size);12 μm (particle size)	Improve dispersibility, coloring strength, and bioavailability	De Paz et al. (2012)
	Maltodextrin	d-Limonene	0.2–1.2 μm (emulsion droplet size); 21–53 μm (dried particle size)	Increase the retention and stability during process	Jafari et al. (2007b)
	Maltodextrin	Fish oil	0.21–5.9 μm (droplet size); 25–41 μm (particle size)	Minimize the unencapsulated oil at the surface and maximize the encapsulation efficiency	Jafari et al. (2008)
Freeze-drying	β-Cyclodextrin, PCL	Fish oil	183–714 nm	Prevent oxidation and mask the odor	Choi et al. (2010)
	PCL	Fish oil	200–350 nm	Increase the oxidative stability and encapsulation efficiency	Bejrapha et al. (2010)
	PCL and gelatin	Capsicum oleoresin	152 nm	Improve the stability	Nakagawa et al. (2011)
	PCL	Capsicum oleoresin	163–1,984 nm	Study the effect of excipients on the stability and particle size of nanocapsules	Bejrapha et al. (2011)
	PCL	Capsicum oleoresin	320–460 nm	Extend the shelf-life and minimize environmental stress	Surassamo et al. (2010)
	Chitosan and zein	α-Tocopherol	200–800 nm	Improve stability and protect from environmental factors	Luo et al. (2011)
	Dioctyl sodium sulfosuccinate	Curcuminoids	450 nm	Improve the stability	Tiyaboonchai et al. (2007)
	Chitosan and sodium tripolyphosphate	Catechin	163 and 165 nm	Protect catechin from degradation	Dube et al. (2010)
	PLGA	Curcumin	264 nm	Improve stability and bioavailability	Shaikh et al. (2009)
	O-Carboxymethyl chitosan	Curcumin	150 nm	Higher anticancer activity	Anitha et al. 2011
	Chitosan-g-poly(N-vinylcaprolactam)	Curcumin	180–220 nm	Higher bioavailability	Rejinold et al. (2011)

PCL poly(ε-caprolactone), PLGA poly(lactide-co-glycolide)

6.1   Spray Drying

Spray drying is a widely used industrial process for the continuous production of dry powders. Spray drying is the process of transforming a feed (solution or suspension) from a fluid into a dried particulate form by spraying the feed into a hot drying medium. The spray drying yields fine particles with less processing time and more economical unit operation (Masters 1991). Due to its continuous production of dry powders with low moisture content, it is widely used for industrial processes (Anandharamakrishnan et al. 2007; Kuriakose and Anandharamakrishnan 2010). Moreover, it is a well-established technique in the food industry and has been widely used for encapsulation for the past few decades.

6.2   Principles of Spray Drying

Spray drying involves three stages of operation: (1) atomization of liquid feed into a spray chamber, (2) contact between the spray and the drying medium, and (3) separation of dried products from the air stream (as shown in Fig. 6.1).

Atomization is a process in which the bulk liquid breaks up into a large number of small droplets. The choice of atomizer is most important in achieving economic production of high quality products (Fellows 1998). The different types of atomizer (Masters 1991) are:

Centrifugal or rotary atomizer: Liquid is fed to the center of a rotating wheel with a peripheral velocity of 90–200 m/s. Droplets are produced typically in the range of 30–120 μm diameter. The size of droplets produced from the nozzle varies directly with the feed rate and feed viscosity, and inversely with wheel speed and wheel diameter.

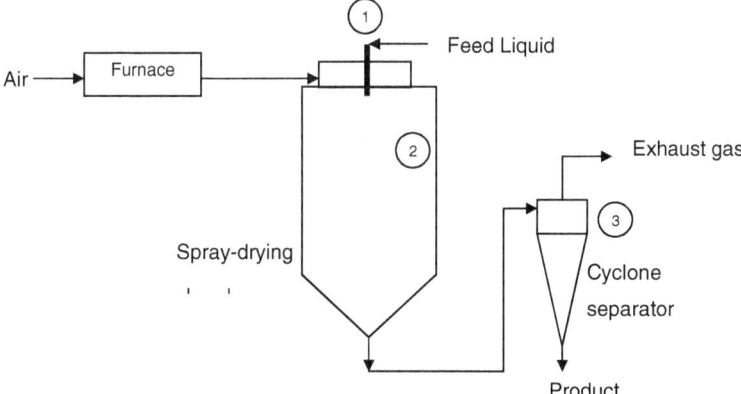

Fig. 6.1 The process stages of spray drying: *1* atomization of liquid feed; *2* contact between the spray and the drying medium; and *3* separation of dried products from the air stream

Pressure nozzle atomizer: Liquid is forced at 700–2,000 kPa pressure through a small aperture. The size of droplets is typically in the range of 120–250 μm. The droplet size produced from the nozzle varies directly with feed rate and feed viscosity, and inversely with pressure.

Two-fluid nozzle atomizer: Compressed air creates a shear field, which atomizes the liquid and produces a wide range of droplet sizes.

During spray–air contact, droplets usually meet hot air in the spraying chamber either in co-current flow or counter-current flow. In co-current flow, the product and drying medium pass through the dryer in the same direction and the arrangement is suitable for the drying of heat-sensitive materials. The advantages of co-current flow processes are rapid spray evaporation, shorter evaporation time, and less thermal degradation of the products (Masters 1991). In contrast, in the counter-current configuration, the product and drying medium enter at opposite ends of the drying chamber. The outlet product temperature is higher than the exhaust air temperature and is almost at the feed air temperature with which it is in contact. This type of arrangement is used for non-heat-sensitive products only. In another type, called mixed flow, the dryer design incorporates both co-current flow and counter-current flow. This type of arrangement is used for drying of coarse free-flowing powders, but the drawback is that the temperature of the product is high.

The dry powder is collected at the base of the dryer and removed by a screw conveyor or a pneumatic system with a cyclone separator. Other methods for collecting the dry powder are bag filters and electrostatic precipitators (Fellows 1998). The selection of equipment depends on the operating conditions such as particle size, shape, bulk density, and powder outlet position. Nanosized particle are separated by electrostatic precipitators or Teflon-coated scrubbers.

Spray drying is also used to encapsulate a wide range of food ingredients such as flavors, vitamins, minerals, colors, fats, and oils in order to protect them from their surrounding environment and to extend shelf-life stability during storage (Pillai et al. 2012). It can be considered a good microencapsulation technique. Encapsulation is achieved by dissolving, emulsifying, or dispersing the core substance in an aqueous solution of carrier material, followed by atomization and spraying of the mixture into a hot chamber (Zuidam and Shimoni 2010). However, the wall material has to be soluble in water and some of the available wall materials are gum acacia, maltodextrins, modified starch, polysaccharides (alginate, carboxymethylcellulose, guar gum), and proteins (whey proteins, soy proteins, and sodium caseinate).

Jafari et al. (2007b, 2008) encapsulated oil through production of nanoemulsions along with spray drying, using modified starch (Hi-Cap) and whey protein concentrate as wall material. In both the studies, only the emulsion droplets were found to be nanosized (200–800 nm) but they were converted to micron size (above 20 μm) during spray drying. Ferreira et al. (2007) encapsulated catechin in a carbohydrate matrix using homogenization followed by spray drying at an inlet temperature of 150–190 °C. The nanocapsules were spherical with a smooth surface and a diameter of 80 nm. The zeta potential of the nanocapsules was highly negative, and that contributed to their stabilization. The release of catechins from the nanocapsules was

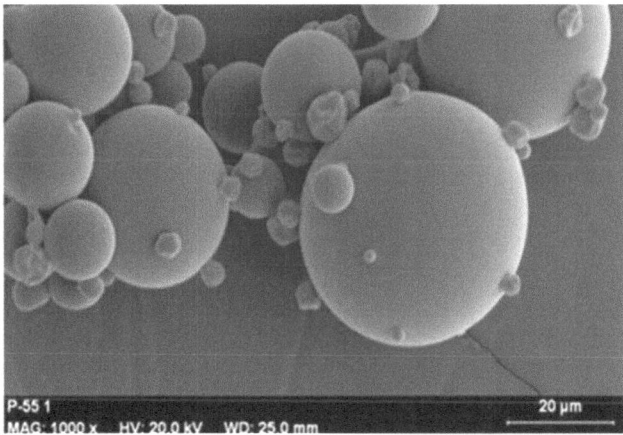

Fig. 6.2 Scanning electron micrograph of particles (nanosuspensions of β-carotene encapsulated using modified *n*-octenyl succinate-starch) obtained after spray drying (De Paz et al. 2012)

observed at pH 5 after heating the solution to a temperature above 80 °C. The stability of the nanocapsules was not affected by decreasing the pH as the catechin content was same at pH 3 and 5. Moreover, encapsulation of catechins prevented them from oxidizing and may improve bioavailability. Similarly, De Paz et al. (2012) encapsulated β-carotene by formulating nanosuspensions using modified *n*-octenyl succinate-starch through emulsification evaporation and spray drying techniques. The nanosuspensions were produced with various experimental operating conditions and resulted in high encapsulation efficiency (65–90%) and antioxidant activity, with particle sizes in the range of 300–600 nm. However, particles collected after spray drying were around 12 μm in diameter (Fig. 6.2). Moreover, it was reported that process parameters such as concentration of modified starch and the organic solvent/water flow ratio have the most influence on the product properties.

Recently, Liang et al. (2013) encapsulated β-carotene through production of nanoemulsions along with spray drying, using modified starch (Hi-Cap) as wall material. The nanoemulsion had mean droplet size of about 114–160 nm. However, these nanoemulsions were converted to micron-sized powders after spray drying. The powders showed a good dissolution in water and the reconstituted emulsions had similar particle sizes to the fresh nanoemulsions, suggesting that the spray drying process did not affect the characteristics of the nanoemulsions. The influence of relative humidity on the storage stability of β-carotene powders at 25 °C for a storage period of 30 days was analyzed. The β-carotene degradation profiles over time were found to be closely related to the film property of the matrix, moisture sorption property, and glass transition temperature of the powder. Moreover, it was found that modified starches with lower film oxygen permeability had a higher retention of β-carotene during storage. The glass transition temperature of powder in different relative humidities also affected the rate of β-carotene degradation.

In contrast to freeze-drying, spray drying is more economical and faster and is a single-step drying method. It yields uniformly spherical particles that offer complete protection of the core material on encapsulation. On the other hand, spray drying yields particles of micron size on drying the nanoemulsions and nanosuspensions. However, the core material encapsulated inside the matrix of micron-sized particles was found to be in the nanosize range (nanosuspensions and nanoemulsions) and, hence, Jafari et al. (2007b, 2008) considered spray drying to be a nanoparticle encapsulation technique. Moreover, nanoencapsulation by spray drying depends on other nanoencapsulation techniques (like emulsification) prior to spray drying. Therefore, the conventional spray drying technique itself may not be considered to be a nanoencapsulation technique. However, in spray drying it is possible to control particle size and morphology by varying process parameters and formulations (Anandharamakrishnan et al. 2007, 2008).

6.3   Freeze-Drying

Freeze-drying or lyophilization is a drying method in which water in a solution or suspension is crystallized at low temperatures and sublimed from the solid state directly into the vapor phase (Anandharamakrishnan et al. 2010). The quality of freeze-dried products is very high in comparison with that of products dehydrated using other techniques, due to the prevention of heat damage (King 1970). A freeze-drying process provides a product that is stable (in dry form), rapidly soluble (large surface area), and elegant (uniformly colored cake). The freeze-drying process consists of four steps as follows:

Freezing: A substance is frozen to form crystallized ice. The crystallization depends on the cooling rate, initial concentration, and end temperature of cooling. Most food components remain in an amorphous, glassy state (i.e., do not crystallize), but the water component does crystallize.

Primary drying: Ice is removed by sublimation at low temperature and low pressure. Sublimation occurs at the interface between the frozen and dry material and this starts at the ice surface.

Secondary drying: Unfrozen water is removed by desorption. This step typically takes one third of the drying time. The final moisture content for food stuffs is 2–10% and for biological products is 0.1–3%.

Final treatment: The drying chamber is filled with an inert gas (nitrogen for foodstuffs, argon for biological products) for preserving the products after drying.

During the freezing of foods, the temperature decreases and water is removed from the food in the form of ice, and the solutes present in the unfrozen products are freeze-concentrated. Hindmarsh et al. (2003) measured the temperatures when freezing single-distilled water droplets on a thermocouple junction and recorded the images of freezing droplets using a high magnification video camera. They observed an expanded recalescence stage.

A problem in the primary drying step is that "collapse" can occur when the viscosity of the structural material is reduced to a level at which it cannot support its weight against gravity (Bhandari et al. 1997). The term "collapse" has been used to describe loss of structure, reduction in pore size, and volumetric shrinkage in dried food materials (Levi and Karel 1995), resulting from a time-, temperature-, and moisture-dependent viscous flow that results in loss of structure. The collapse temperature (T_c) decreases with decreasing molecular weight and also decreases with increasing moisture content (Oetjen 1999). T_c can be related to the glass transition temperature (T_g). When the temperature is higher than T_g, the amorphous matrix viscosity decreases.

Freeze-drying is a highly stabilizing process and is generally applied to enhance the physicochemical stability of the nanoparticles to achieve an acceptable product, especially in case of unfavorable storage conditions. However, energy intensiveness, long processing time (more than 20 h) and an open porous structure are the main drawbacks of freeze-drying (Singh and Heldman 2009). Nevertheless, freeze-drying is normally used for separation of nanoparticles (i.e., removal of the water from the substances) produced by other nanoencapsulation techniques. During freeze-drying, pores are formed due to ice sublimation. Hence, this process is not purely encapsulation because active food ingredients are exposed to the atmosphere due to the presence of pores on the particle surface. Therefore, it is difficult to use any release mechanism such as diffusion or erosion techniques. Currently, freeze-drying is widely used to remove the water from nanocapsules without changing their structure and shape. However, spray-freeze-drying may be an effective alternative to conventional freeze-drying in terms of reducing the pore size and drying time (Anandharamakrishnan et al. 2010 and Ezhilarasi et al. 2013).

Shaikh et al. (2009) encapsulated curcumin by an emulsion–diffusion–evaporation method along with freeze-drying. The nanocapsules had a mean diameter of about 264 nm and 77 % entrapment efficiency at 15 % loading. The curcumin-loaded nanoparticles were reported to be stable during a 3-month stability study period. The in vivo pharmacokinetics revealed that curcumin entrapped in nanoparticles demonstrated at least a nine fold increase in oral bioavailability compared to curcumin administered with piperine as absorption enhancer. Similarly, Anitha et al. (2011) developed curcumin-loaded O-carboxymethyl chitosan nanoparticles by an ionic crosslinking reaction along with freeze-drying. The nanoparticles were found to be spherical with 150 nm mean diameter and 87 % entrapment efficiency. The in vitro release profile indicated slow, controlled and sustained release of curcumin from the nanocapsules as well as enzyme-triggered degradation and release in the presence of lysozyme. The nanocapsules also exhibited toxicity towards cancer cells and nontoxicity towards normal cells. Similarly, Rejinold et al. (2011) encapsulated curcumin with biodegradable thermoresponsive chitosan-g-poly(N-vinylcaprolactam) by a simple ionic crosslinking method using tripolyphosphate along with freeze-drying. The nanocapsules were in spherical in shape with a size range of 180–220 nm. Curcumin-loaded nanoparticles showed specific toxicity to cancer cells.

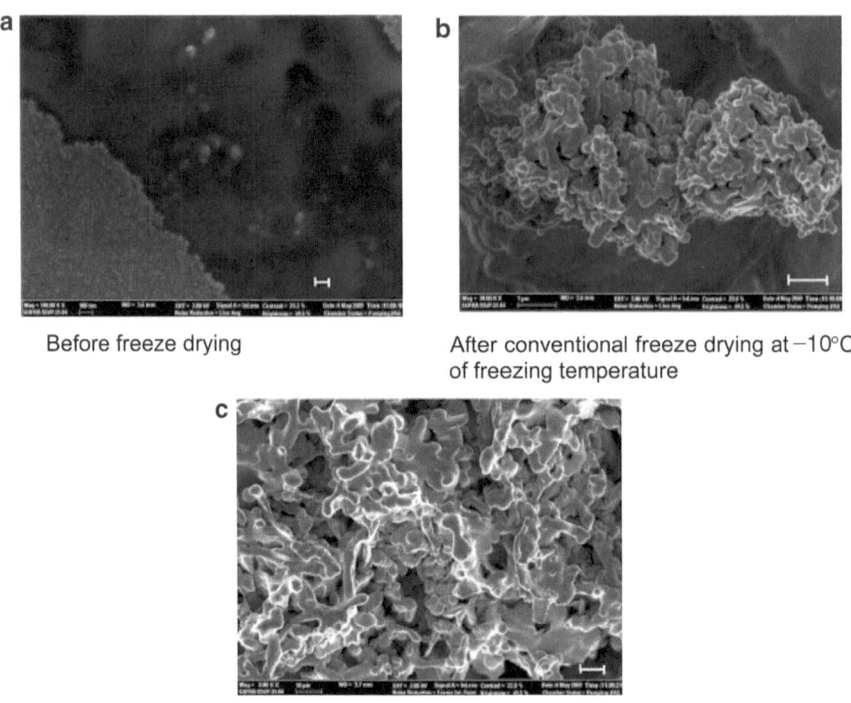

Before freeze drying   After conventional freeze drying at −10°C
                                             of freezing temperature

After vacuum freeze drying at −30°C of freezing temperature

Fig. 6.3 Scanning electron micrograph of fish oil-loaded nanocapsules (Bejrapha et al. 2010):
(**a**) before freeze-drying, (**b**) after conventional freeze-drying at −10 °C freezing temperature, and
(**c**) after vacuum freeze-drying at −30 °C freezing temperature

Fish oil (FO) was encapsulated using β-cyclodextrin (β-CD) by a self-aggregation
method and also using poly(ε-caprolactone) (PCL) via an emulsion diffusion
method, followed by freeze-drying (Choi et al. 2010). The mean particle size of
β-CD–FO was reported to be 250–700 nm and that of PCL:FO particles was less
than 200 nm. It was found that PCL:FO had a higher fish oil loading, lower fish oil
leakage, and higher encapsulation efficiency (99 %) than β-CD–FO (84–87 %). The
storage stabilities of freeze-dried β-CD–FO complexes at 10:20 (w/w) mixing ratio
at various relative humidities retained 97 % of fish oil within the particles for 3 days.
Moreover, PCL more efficiently retarded the release of fish oil in liquid or powder
form due to its water insolubility, despite the fact that particles were broken by
freeze-drying. Recently, Bejrapha et al. (2010) compared the effects of vacuum
freeze-drying (vacuum pressurized freezing and drying) and conventional freeze-
drying (atmospheric pressurized freezing and drying) processes on the stability of
fish oil-loaded nanocapsules, encapsulated in PCL. The nanocapsules were reported

to be below 360 nm in size and were aggregated (Fig. 6.3) with thin membrane layers surrounding the surface of the fish oil. The encapsulation efficiency of conventional freeze-drying was greater than that of vacuum freeze-drying, except at a freezing temperature of −30 °C. In addition, the authors revealed that the vacuum-freezing process may affect the fragility of the PCL membrane due to its lower encapsulation efficiency and aggregation of particles. It was also found that the conventional freeze-drying process was more effective than vacuum freeze-drying in improving the oxidative stability of fish oil-loaded nanocapsules. Vacuum freeze-dried nanoparticles did not endure the stress of vacuum freezing, as shown by breaks in weak and thin membranes, resulting in oil release from their interiors.

Capsicum oleoresin was encapsulated in PCL using the emulsion diffusion method, which involves the formation of an emulsion between water-miscible solvent containing drug and polymer aqueous phase (Surassamo et al. 2010). Addition of water to the system causes solvent to diffuse to the external phase, resulting in formation of nanospheres (Quintanar-Guerrero et al. 1998); this is followed by freeze-drying. By varying the concentration of surfactant (Pluronic F68), the process parameters were optimized. The obtained nanoemulsions were in the size range 320–460 nm. The size of nanocapsule particles decreased on increasing the emulsifier and surfactant concentrations. Similarly, using a modified emulsion–diffusion method combined with freeze-drying, Bejrapha et al. (2011) produced capsicum oleoresin-loaded nanocapsules with PCL. Various freezing temperatures such as −40 °C, −20 °C, and −15 °C were applied to study the effects of cooling temperature on the properties of the capsicum oleoresin-loaded nanocapsules. The effects of excipients such as gelatin and κ-carrageenan on the stability of nanocapsules during freeze-thawing and freeze-drying procedures were studied. The results clearly showed that a relatively high freezing temperature (−15 °C) had an effect on the maintenance of nanocapsule size after freeze-thawing and freeze-drying.

Capsicum oleoresin was encapsulated in PCL, stabilized with gelatin through emulsion-diffusion followed by freeze-drying, and its dispersibility studied (Nakagawa et al. 2011). The mean diameter of the nanocapsules was below 200 nm. Moreover, the freeze-dried capsules had different dispersion characteristics at different positions in the dried bulk sample. This heterogeneity was dependent on the cooling program used during the processing. The freezing front velocity and the viscosity of the solution at/around the freezing front were estimated by using the simulated thermal profiles. A correlation was found between the dispersibility of the freeze-dried nanocapsule–gelatin suspension, the freezing front velocity, and the viscosity at/around the freezing front. It was suggested that the gel network formation of nanocapsule gelatin would be advantageous to avoid unfavorable denaturation and to produce excellent dispersion characteristics in the dried matrix.

Abdelwahed et al. (2006b) studied the freeze-drying of PCL nanocapsules encapsulating miglyol 829 oil prepared by the emulsion-diffusion method and stabilized by PVA. Different parameters were tested throughout the freeze-thawing study, including PVA and PCL concentration, cooling rate, cryoprotectant concentrations (sucrose and polyvinyl pyrolidone), nature of encapsulated oil, and nanocapsule purification. The type of cryoprotectants had practically negligible effects on the size and the rehydration of freeze-dried nanocapsules, and the annealing process accelerated sublimation with the conservation of nanocapsule size. Dube et al. (2010) encapsulated (+)-catechin and (−)-epigallocatechin gallate (EGCG) in chitosan–tripolyphosphate by an ionic gelation method and sonication along with freeze-drying. The effectiveness of nanoencapsulation was compared with the method of adding reducing agents such as ascorbic acid, dithiothreitol, or tris(2-carboxyethyl)phosphine (TCEP) for their potential to protect catechin and EGCG from degradation. The nanocapsules had a mean particle size of less than 200 nm. Encapsulation in chitosan–tripolyphosphate had effectively protected the catechins. It took 8 and 24 h for the non-encapsulated and encapsulated (+)-catechins, respectively, to degrade to 50 % of their initial levels, and the corresponding values for the nonencapsulated and encapsulated (−)-epigallocatechin gallate were 10 and 40 min, respectively. Nanoencapsulation of catechin and EGCG provided better protection than reducing agents TCEP and ascorbic acid. Recently, Luo et al. (2011) encapsulated tocopherol in zein and a zein/chitosan complex using a freeze-drying technique. The physicochemical and structural analysis showed that the electrostatic interactions and hydrogen bonds were major forces responsible for complex formation. The scanning electron microscopy study revealed the spherical nature of the nanocapsules and the smooth surface of the complex. The particle size of the complex varied from 200 to 800 nm and encapsulation efficiency was 77–87 %.

Freeze-drying can improve the stability of core compound against degradation and result in better encapsulation efficiency of above 70 % (Ezhilarasi et al. 2013). However, the characteristics of the final freeze-dried nanoparticles depend on a suitable high-energy emulsification technique and other encapsulation techniques to break down the droplets into nanosized form. Moreover, freeze-drying requires cryoprotectants such as sucrose, trehalose, or mannitol to conserve the particle size and to avoid aggregation during freeze-drying. The various freezing temperatures are also reported to affect the nanocapsule size (Ezhilarasi et al. 2013). Thus, freeze-drying seems to be an efficient drying technique for stabilizing nanocapsules and is capable of retaining the particle size in the nanometric range, even after drying.

Chapter 7
Applications of Food-Grade Nanoemulsions

Food-grade nanoemulsions constitute one of the most promising systems for improving the solubility, bioavailability, stability, and functionality of many bioactive compounds. These improved properties of nanoemulsions can be exploited for many novel and technological innovations that find various industrial applications. Furthermore, there are challenges such as suitable processing operations and facilities to scale-up for industrial production that need to be overcome for wide utilization of nanoemulsions. This chapter reviews the various applications of nanoemulsions.

7.1 Application of Food-Grade Nanoemulsions

The major applications of food-grade nanoemulsions are currently as antimicrobial agents and for transdermal delivery. These, as well as the bioavailability and bioefficacy of the bioactive compounds, are discussed in detail in the following sections.

7.1.1 Antimicrobial Activity

The use of bioactive compounds (essential oils) as natural antimicrobial agents in extending the shelf life of foods has been well established through various research studies. The same bioactive compounds in nanoemulsion form exhibit higher antimicrobial activity than the conventional form due to the smaller droplet size. The discrete droplets of nanoemulsion selectively fuse with bacterial cell walls or viral envelopes, destabilizing the pathogen's lipid envelope and initiating their disruption (Baker et al. 2003). Moreover, the nanoemulsion may potentially increase the passive cellular absorption mechanisms, thus reducing the mass transfer resistances

C. Anandharamakrishnan, *Techniques for Nanoencapsulation of Food Ingredients*,
SpringerBriefs in Food, Health, and Nutrition, DOI 10.1007/978-1-4614-9387-7_7,
© C. Anandharamakrishnan 2014

and increasing antimicrobial activity (Donsi et al 2011a). Donsi et al. (2011a) investigated the antimicrobial activity of nanoemulsions prepared with various essential oils such as carvacrol, limonene, and cinnamaldehyde against three different microorganisms, *Escherichia coli*, *Lactobacillus delbrueckii*, and *Saccharomyces cerevisiae*. The nanoemulsions were stabilized by different emulsifiers (lecithin, pea proteins, sugar ester, and a combination of Tween 20 and glycerol monooleate) and had a mean droplet size of about 100–200 nm. The nanoemulsions based on sugar esters or combination of Tween 20 and glycerol monooleate exhibited higher bactericidal activity over short time scales (2 h) whereas nanoemulsions based on lecithin or pea protein exhibited bactericidal activity over a longer time scale (24 h). The antimicrobial activity of nanoemulsions was correlated to the concentration of active molecules in the aqueous phase and the emulsifier capability. Following this study, Donsi et al. (2011b) investigated the minimal inhibitory concentration (MIC) and minimal bactericidal concentration (MBC) of nanoemulsions prepared with a terpene mixture and D-limonene against similar classes of microorganisms. It was reported that antimicrobial activity increased with the change in the emulsion droplet size as well as with the class of microorganism. Moreover, the high intensity processing during nanoemulsion production may affect the chemical stability of several active compounds. The application of the most efficient antimicrobial nanoemulsions was tested in pear and orange juices.

Joe et al. (2012a) prepared nanoemulsions using selected cooking oils such as sunflower, castor, coconut, groundnut, and sesame oils using a biosurfactant (Surfactin). The Surfactin-based sunflower oil nanoemulsion (AUSN) was highly transparent with a particle size of 72 nm and survived thermodynamic stability tests with low surface tension, viscosity, and density. Moreover, AUSN showed the highest activity against *Salmonella typhi*, followed by *Listeria monocytogenes*, and *Staphylococcus aureus*. It also demonstrated high fungicidal activity against *Rhizopus nigricans*, followed by *Aspergillus niger* and *Penicillium* sp. and a greater sporicidal activity against *Bacillus cereus* and *Bacillus circulans*. In situ evaluation of AUSN for antimicrobial activity on food products such as raw chicken, apple juice, milk, and mixed vegetables showed a significant reduction in the bacterial and fungal populations of these products. Continuing this work, Joe et al. (2012b) studied the influence of AUSN on the microbiological, proximal, chemical, and sensory qualities of Indo-Pacific king mackerel (*Scomberomorus guttatus*) steaks stored at 20 °C over a time period of 72 h. AUSN treatment significantly decreased the values of chemical indicators of spoilage throughout the storage period, and organoleptic evaluation exhibited an extension of shelf life up to 48 h compared with control and antibiotic-treated samples. Likewise, Sugumar et al. (2012) studied antimicrobial activity of eucalyptus oil nanoemulsion (16 nm) against three different microorganisms (*B. cereus*, *S. aureus,* and *E. coli*). Nanoemulsions showed 100 % bactericidal activity at tenfold dilution. Recently, Karthikeyan et al. (2012) investigated the MIC and MBC of soybean oil nanoemulsion (308 nm) against *Streptococcus mutans*, *Lactobacillus casei, Actinomyces viscosus, Candida albicans*, and mixed culture. However, higher MIC was observed for *A. viscosus* and *S. mutans*.

7.1.2 Transdermal Delivery

Nanoemulsions have great potential for transdermal delivery of drugs and bioactive compounds. The rate and extent of active compound/drug penetration into different layers of skin and into systemic circulation are governed by the physicochemical properties of the drug, such as drug solubility in the vehicle, relative solubility in the vehicle and skin (partition coefficient), molecular size, and formulation characteristics. The concentration gradient drives the passive permeation of drug molecules through the skin (Venuganti and Perumal 2009). Hence, the nanoemulsions having smaller droplet size can improve the active compound/drug solubility, which in turn enhances skin penetration and permeability. Zhou et al. (2010) prepared a lecithin nanoemulsion of droplet size 58–92 nm using the high-pressure homogenization technique. In an in vivo study, cream incorporated with lecithin nanoemulsions exhibited 2.5-fold higher skin hydration capacity at 10 % concentration. Moreover, lecithin nanoemulsions loaded with Nile Red dye showed improved penetrability into the dermis layer on the abdominal skin of rat.

Similarly, Shakeel and Ramadan (2010) produced a caffeine nanoemulsion and performed in vitro skin permeation studies on Franz diffusion cells using rat skin as permeation membrane. Caffeine nanoemulsions exhibited significant increase in skin permeability as compared to an aqueous solution of caffeine. The enhancement ratio was found to be 17 in an optimized nanoformulation compared with other formulations. Recently, Nam et al. (2012) prepared nanoemulsions using tocopheryl acetate along with a mixture of unsaturated phospholipids and polyethylene oxide-*block*-poly(ε-caprolactone) (PEO-*b*-PCL). The effects of the lipid–polymer composition on the size and surface charge of the nanoemulsions, microviscosity of the interfacial layer, and skin absorption of tocopheryl acetate were investigated. The amount of tocopheryl acetate absorbed by the skin increased with an increase in lipid-to-polymer ratio, as determined using guinea pig skin loaded in a Franz-type diffusion cell. A mixture of unsaturated phospholipids and PEO-*b*-PCL in 8:2 ratio exhibited the most efficient delivery of tocopheryl acetate into the skin.

7.1.3 Bioavailability and Bioefficacy

Bioavailability is a complex process involving several different stages: liberation, absorption, distribution, metabolism, and elimination. Bioavailability is a key step in ensuring bioefficacy of active food compounds and oral drugs (Rein et al. 2012). A nanoemulsion is a good candidate for delivery of various bioactive compounds due to its ability to improve bioactive solubilization and absorption in the gastrointestinal tract, caused by surfactant-induced permeability changes. The bioactivity of a component, which is a measure of its specific biological affect, is determined by the fraction available for absorption. Improving the bioavailability of bioactive food compounds can improve their bioefficacy (Rein et al. 2012). Hatanaka et al. (2008)

reported the 1.7-fold higher bioavailability of coenzyme Q10 nanoemulsions (60 nm) than its crystalline form in an in vivo study.

Similarly, Kuo et al. (2008) produced nanoemulsions of an antioxidant synergy formulation (ASF), containing delta-, alpha- and gamma-tocopherol and analyzed their influence on anti-inflammatory activity and bioavailability in mice. The ASF-nanoemulsions exhibited enhanced anti-inflammatory properties compared with their suspensions, and nanoemulsions of gamma-tocopherol showed a significant effect. The bioavailability of nanoemulsions with gamma- and delta-tocopherol was enhanced (2.2- and 2.4-fold, respectively) in comparison to the suspensions. Similarly, Wang et al. (2008b) reported enhanced anti-inflammatory activity of the curcumin nanoemulsions in the mouse ear inflammation model. The nanoemulsions produced through high-speed homogenization (619 nm) and high-pressure homogenization (80 nm) exhibited 43 % and 85 % inhibition, respectively, of mouse ear edema induced by 12-O-tetradecanoylphorbol-13-acetate.

Vishwanathan et al. (2009) analyzed the bioavailability of lutein nanoemulsions (150 nm) in human subjects. The subjects consumed a lutein supplement pill or a lutein nanoemulsion added to orange juice (6 mg/day in study 1 and 2 mg/day in study 2) for 1 week. The mean serum lutein concentrations increased by 104 % and 167 % after consuming the 6 mg supplement or nanoemulsion, respectively (study 1). Similarly, the consumption of 2 mg lutein supplement or nanoemulsion increased serum lutein concentrations by 37 % and 75 %, respectively (study 2). Despite the fact that there was lutein loss during nanoemulsion preparation, subjects consuming nanoemulsions showed greater serum lutein concentrations compared with subjects receiving the lutein supplement. Based on the two studies, it was concluded that lutein nanoemulsions had significantly greater bioavailability than the supplement in pill form.

Recently, Ragelle et al. (2012) investigated the antitumor activity and bioavailability of a nanoemulsion produced using fisetin (3,3′,4′,7-tetrahydroxyflavone), a natural flavonoid. The nanoemulsion had mean droplet diameter of 153 nm and was stable at 4 °C for 30 days. Pharmacokinetic studies in mice reported that fisetin nanoemulsion had 24-fold increased relative bioavailability than free fisetin when administered intraperitoneally. Additionally, fisetin nanoemulsion exhibited antitumor activity in mice at lower doses (37 mg/kg) than free fisetin (223 mg/kg). Likewise, Anuchapreeda et al. (2012) formulated curcumin lipid nanoemulsions (47–55 nm droplet size) by a modified thin-film hydration method using soybean oil, hydrogenated L-α-phosphatidylcholine, and co-surfactants. The nanoemulsions were stable for 60 days at 4 °C with respect to particle size. Curcumin nanoemulsions showed more efficient anticancer activity than curcumin solution in cytotoxicity studies using B16F10 and leukemic cell lines.

Chapter 8
Characterization of Nanoparticles

Synthesis of nanoparticles using different techniques of encapsulation has attracted a lot of interest these days and the incorporation of nanoparticles into food items has made food more "functional." But, this field of research is still under trial and is not yet knocking on the door of industry. There are some limitations that are hindering the path of revolution. Before the incorporation of nanoparticles into any food product or drug, it is necessary to characterize the behavior of the miniature product. This chapter reviews the various techniques for characterization of nanoparticles.

8.1 Analysis of Particle Size and Morphology

To analyze the size of micro- and nanoscale particles, dynamic light scattering (DLS) and electron microscopy are preferred. Two types of electron microscopy are mainly used: scanning electron microscopy (SEM) and transmission electron microscopy (TEM). Electron microscopy offers better resolution than the optical microscopy. Surface morphology of the particles can be analyzed by atomic force microscopy (AFM).

8.1.1 Dynamic Light Scattering

The most common tool for analyzing the size of nanoparticles is dynamic light scattering (DLS). The basic principle of this method is to observe the movement of particles and measure their Brownian motion. The velocity of the particles is inversely proportional to the size of the particle and is called translational diffusion coefficient (Quintanilla-Carvajal et al. 2010). DLS mainly calculates the fluctuation

C. Anandharamakrishnan, *Techniques for Nanoencapsulation of Food Ingredients*,
SpringerBriefs in Food, Health, and Nutrition, DOI 10.1007/978-1-4614-9387-7_8,
© C. Anandharamakrishnan 2014

occurring in the path of light passing through the sample due to Brownian motion of the particles (Nobbmann et al. 2007). The DLS method is very sensitive and able to measure particle size down to the 0.001 µm level, which cannot be measured with any other method. Its major advantages are cost-effectiveness and quick results. Moreover, the properties of the solution are not deteriorated by the beam. However, this technique is very crude, performs indirect measurements, and can only be used for liquids and gels.

8.1.2 Scanning Electron Microscopy

In scanning electron microscopy (SEM), an electron beam from a tungsten source or lanthanum hexaboride thermionic emitters is allowed to strike the surface of the sample for its visualization. The filaments emit the electrons when heated within the temperature range of 2,000–2,700 K. An electron gun generates the electron beam with energy of 0.1–30 keV. The electron beam passes through the influence of an electromagnetic field generated by lenses, which forces the electron beam to strike the surface of the sample (Lawes 1987). After striking the surface of sample, the electron beam is diffracted in different directions and generates a number of signals that can be imaged on the screen. The signal-generating electrons are mainly of two categories: secondary electrons (SE) and backscattered electrons (BSE) that generate the SEM image of the sample. The image is generated when a positive voltage is applied to the collector screen in the front of the detector. As the electron hits the inner shell electron and knocks it out, the electron from the outer shell drops to lower energy levels and allows the low energy Auger electrons to be expelled. This technique is called Auger electron spectroscopy (Lawes 1987). SEM is mainly used for micron-sized particles and can also occasionally be used for nanoparticles. Moreover, SEM helps to reveal the high degree of dispersion and uniformity of the metallic nanoparticles (Herrera and Sakulchaicharoen 2009).

8.1.3 Transmission Electron Microscopy

The basic principle of transmission electron microscopy (TEM) is derived from the de Broglie wavelength. A limit is set for the wavelength, which gives maximum resolution when a beam of electrons penetrate the sample. The basic difference between the SEM and TEM is that former scans the surface of sample whereas latter penetrates through the sample and generates an image of the particle by interacting with the electrons and is best suited to analyze nanosized particles. TEM provides better resolution than SEM because the energy of the electron beam is higher in TEM (Williams 1996).

8.1.4   Atomic Force Microscopy

Atomic force microscopy (AFM) is mainly used to study the surface morphology of a particle in the x, y, and z directions. This technique has created a revolution in the research field. A cantilever with a sharp tip is allowed to move over the surface of the sample and bends according to the response given to force applied between the tip and sample. AFM has been observed to overcome the limitations of the microscopy imaging techniques (SEM and TEM), as they can image only conducting surfaces. AFM can also be used to study the structural characterization of bioactive molecules and the morphology, size, and process dynamics of lipid nanocapsules (Luykx et al. 2008). However exposure to the rough surfaces when the tip is in direct contact with surface, will impose difficulties.

8.2   X-Ray Diffraction

X-ray diffraction (XRD) is mainly used for determining the basic crystal structure of the particles. The interatomic distance of any crystalline particle is nearly 2–3 Å, which is found to be compatible with the electromagnetic wavelength of the X-rays. A diffraction pattern is observed when a beam of X-rays strikes the different layers of the crystal at a specific angle; consequently, a constructive and destructive interference is observed. Three types of detectors are mainly used to capture the diffracted electrons: photographic film, imaging plate, and charge-coupled device are used to produce a digital image from the diffracted electrons.

Various studies have demonstrated that XRD is a suitable technique for investigating the structure of nanoparticles. Sathishkumar et al. (2009) synthesized silver nanoparticles and confirmed their crystalline structure using TEM and XRD analysis. XRD has also stretched its arms to characterize lipid nanocapsules as well as to study albumin-zinc nanoparticles, albumin-protamine-oligonucleotide nanoparticles, and lactoglobulin aggregates (Luykx et al. 2008).

8.3   Zeta Potential

Zeta potential is the measure of overall charge a particle acquires in a specific medium and gives an indication of the potential stability of a colloidal system (Tiede et al. 2008). It also explains the electrostatic interaction and mobility of a colloidal solution. Electrostatic repulsion interaction is used to measure and control the stability of the solution. It explains the reasons for the occurrence of dispersion, aggregation, or flocculation and can be used to improve the conditions of the colloidal solution (Weiner et al. 1993). Van der Waals forces are present over a short range and dominate the electric charge and cause aggregation in solution. This mainly occurs when the particles in the solution are nonpolar or hydrophobic in nature. An extra charged attachment is required in that case to stabilize the solution.

Chapter 9
Safety and Regulations: Current Scenario and Scope

Today's consumers are conscious about what they eat. "Safety" is the buzzword in the current food scenario. Food safety is an area of major concern today and there are many demands on the food production system. Delivering safe food is the responsibility of all the stakeholders in the food processing chain, including researchers and manufacturers. As with any new technology, nanofoods are also expected to face this safety challenge and gain acceptance before they find a place on the consumer's shelf. To date, there is no clear implication that nanofoods are either safe or dangerous or that they have harmed human health on ingestion. According to the World Health Organization, it is established that consumers are likely to benefit from nanotechnology, however new data and measurement approaches are needed to ensure the safety of products. Incorporating nanotechnology into food systems should be done with clarity on its health and safety impacts. In this context, the safety considerations of nanofoods are briefly discussed in this chapter.

The food that we consume inherently contains components like proteins and carbohydrates whose molecular size falls in the nanoscale dimension. Food proteins, which are globular particles between tens and hundreds of nanometers in size, are true nanoparticles. Linear polysaccharides with one-dimensional nanostructures are less than 1 nm in thickness, and starch polysaccharides having small 3D crystalline nanostructures are only tens of nanometers in thickness (Chi-Fai et al. 2007) and are not known to pose any safety issues. Hence, it is necessary to understand why the debate on safety implications centers only on the nanoingredients that are deliberately added to foods. From the reports already available on the safety concerns of nanofoods, it can be seen that uncertainties on the use of nanoparticles in food or food contact materials still exist (Cheftel 2011). Furthermore, the in-depth potential risks of nanomaterials to human health and to the environment are still unknown (Dowling 2004). This is because there is little or no scientific information on the effects of nanotechnology applications on human and animal health or on the

C. Anandharamakrishnan, *Techniques for Nanoencapsulation of Food Ingredients*, SpringerBriefs in Food, Health, and Nutrition, DOI 10.1007/978-1-4614-9387-7_9, © C. Anandharamakrishnan 2014

environment (Casabona et al. 2010). Moreover, clear regulatory principles on the use of nanomaterials in foods are lacking. All the above reasons mean that the questions on the safety of nanofoods remain unanswered after several years.

According to Chaudhry and Castle (2011), the concern about consumption of nanofoods is split into three areas, namely, that of least concern, some concern, and major concern. The area of least concern is where processed food that contains non-biopersistent nanomaterials is digested or solubilized in the gastrointestinal tract. Where a food product contains non-biopersistent nanomaterials that are not digested but carried across the gastrointestinal tract is an area of some concern. Other areas of some concern are increased bioavailability of vitamins and minerals and a greater uptake of food colors or preservatives so that uptake exceeds the acceptable daily intake (ADI) and may not always be beneficial for consumer health. The areas of major concern are where foods include insoluble, indigestible, and potentially biopersistent nanoadditives (for example, metals or metal oxides) or functionalized nanomaterials. Such applications may pose a risk of consumer exposure to "hard" nanomaterials, the adsorption, distribution, metabolism, and elimination (ADME) profile and toxicological properties of which are not fully known at present.

Another concern raised by Mukul et al. (2001) was that most nanomaterials used in foods are organic moieties and may contain and carry other foreign substances into the blood through the nutrient delivery system. ZnO is an example of an organic moiety that has a zinc moiety and an oxygen moiety. Powell et al. (2010) also reported that it is possible, under certain conditions, for very small nanoparticles to gain access to the gastrointestinal tissue via paracellular transcytosis across tight junctions of the epithelial cell layer. However, whether there are realistic situations of nanoparticle exposure that lead to significantly abnormal reactive oxygen species and inflammasome activation responses (in vivo) in the gut has not been established.

Thus, the available literature on nanofood safety shows clearly that the negative statements about the risks associated with nanofood consumption are assumptions and lack a strong scientific basis. Still, it is important to bridge the gaps between hypotheses on nanofood safety and the gaps in knowledge through intensive research and risk assessment studies. This necessity has drawn the attention of regulatory authorities to establish more organized rules and regulations for the use of nanofoods.

Many nanotechnology initiatives, commissions, or centers have been launched by governments, academia, and private sectors in the USA, Europe, Japan, and some other countries around the globe to ensure the rapid development and deployment of nanotechnology, promote economic growth, maintain global competitiveness, and improve the innovative capability (Chen et al. 2006b; ETC Group 2005). Up to now, there is no international regulation of nanotechnology or nanoproducts. Only a few government agencies or organizations from different countries have established standards and regulations to define and regulate the use of nanotechnology. The existing regulations on nanotechnology are presented by countries such as the USA, UK, Japan, and mainland China. The US Food and Drug Administration

(FDA) is one of the first government agencies around the world to give a definition for nanotechnology and nanoproducts. Details of organizations and institutions initiated to work on the safety prospects of nanofoods have been listed by Chau et al. (2007).

In a few countries (e.g., the USA), some existing laws, notably the Toxic Substances Control Act; the Occupational Safety and Health Act; the Food, Drug and Cosmetic Act; and the major environmental laws provide some legal basis for regulating nanotechnology. The Codex Alimentarius Commission, jointly established by the Food and Agriculture Organization (FAO) and the World Health Organization (WHO), aims at monitoring and coordinating the food safety and regulation standards formulated by food operators worldwide. The involvement of this premier organization in regulating the use of nanocomponents and nanoscale equipments in foods may bring positive insight into the concept of nanofood safety. The National Nanotechnology Initiative (NNI) set out to describe the uses and applications of nanotechnology in three areas, namely, development of products of dimensions 1–100 nm, creating and using nanostructures that exhibit novel properties because of their small sizes, and the ability to control on the atomic scale (Chau et al. 2007). The above are some of the measures put forth to initiate and bring nanofoods within the boundaries of regulatory bodies.

Even though the above-mentioned initiatives are expected to strengthen the safety and regulatory aspects of nanofoods, a proactive approach is needed to identify the potential areas of improvement. The European Food Safety Authority (EFSA 2009) has issued a draft opinion that there are broad uncertainties over the safe use of nanotechnology for foodstuffs, and more research is recommended. The European Commission carried out a review of nanotechnology regulations adopted in June 2008 and concluded that "Current legislation covers in principle, the potential health, safety and environmental risks in relation to nanomaterials. The protection of health, safety and the environment needs mostly to be enhanced by improving implementation of current legislation." Earlier studies have provided many recommendations with the view of improving nanofood safety. The significant recommendations suggested by different groups are listed in Table 9.1.

Thus it is clear that, even though nanofoods are promisingly expected to deliver their full potential in future years, the safety and regulatory aspects are still at the nascent stage. Walsh et al. (2008) commented on this scenario that the pace of the regulatory process lags far behind the speed of nanotechnology products' commercial introduction. The regulations must be designed as a precautionary measure to promote nanofood safety but not to prevent the innovations expected in the applications of nanotechnology in the food sector. With the nanofood products expected to flood the food market in near future, it is necessary that the ambiguities on the safety of nanofoods be cleared to enable consumers to pick nanofoods from the market shelves with trust and confidence.

Table 9.1 Recommendations on the development of regulations for nanofoods

Recommendation	Author(s)/group
Concerning the changes in the bioactivity, physicochemical property, and functions of nanoscale materials in relation to their size reduction	Chau et al. (2007)
Dividing nanofood products into different categories such as liquid, powder, aerosol, suspension, emulsion, and liposome for proper classification, management, and analysis	
Safety assessment or nanotoxicity studies should be required for nanoscale materials that are chemically modified	
Requirement of food labeling to identify the presence of nanomaterials in products and provide possible particle size range and relevant safety information	
Regulations governing nanomaterial developments, verification of their safety, fate, and how to dispose of them through remediation treatments need to be understood	Neethirajan and Jayas (2010)
Research into the consequences of the ingestion of nanoparticles	International Union of Food Science and Technology (2007)
Research on materials not normally adsorbed, digested, and metabolized on ingestion	
Where the safety data or ADIs for manufactured nanoparticles of food-approved ingredients or additives differs from that of the bulk materials, then there may be a need for selective or distinctive labeling of these nanoproducts	
The toxicokinetic properties of engineered nanomaterials after oral exposure into the human body should be correlated with their physicochemical properties to determine whether these nanomaterials can be categorized on the basis of appropriate dose metrics	European Food Safety Authority (EFSA 2009)

References

Abdelwahed W, Degobert G, Trainmesse S, Fessi H (2006a) Freeze-drying of nanoparticles: formulation, process and storage considerations. Adv Drug Deliver Rev 58(15):1688–1713.

Abdelwahed W, Degobert G, Fess H (2006b) A pilot study of freeze drying of poly(epsilon-caprolactone) nanocapsules stabilized by poly(vinyl alcohol): formulation and process optimization. Int J Pharmaceut 309:178–188.

Acosta E (2009) Bioavailability of nanoparticles in nutrient and nutraceutical delivery. Curr Opin Colloid Interface Sci 14(1):3–15.

Aditya NP, Shim M, Inae L, Lee YJ, Im M-H, Ko S (2013)Curcumin and genistein co-loaded nanostructured lipid carriers: in vitro digestion and anti-prostate cancer activity J Agric Food Chem 61(8): 1878–1883.

Alexander M, Lopez AA, Fang Y, Corredig M (2012) Incorporation of phytosterols in soy phospholipids nanoliposomes: Encapsulation efficiency and stability. LWT – Food Sci Technol 47(2):427–436.

Anand P, Nair HB, Sung B, Kunnumakkara AB, Yadav VR, Tekmal RR, Aggarwal BB (2010) Design of curcumin-loaded PLGA nanoparticles formulation with enhanced cellular uptake, and increased bioactivity in vitro and superior bioavailability in vivo. Biochem Pharmacol 79(3):330–338.

Anandharamakrishnan C, Rielly CD, Stapley AGF (2007) Effects of process variables on denaturation of whey protein during spray drying. Dry Technol 25(5):799–807.

Anandharamakrishnan C, Rielly CD, Stapley AGF (2010) Spray- freeze- drying of whey proteins at sub atmospheric pressures. Dairy Sci Technol 90(2–3):321–334.

Anandharamakrishnan C, Rielly CD, Stapley AGF (2008) Loss of solubility of α-lactalbumin and β-lactoglobulin during spray drying of whey proteins. LWT- Food Sci Tech 41(2): 270–277.

Anarjan N, Mirhosseini H, Baharin BS, Tan CP (2011) Effect of processing conditions on physicochemical properties of sodium caseinate-stabilized astaxanthinnanodispersions. LWT – Food Sci Tech 44(7):1658–1665.

Anitha A, Maya S, Deepa N, Chennazhi KP, Nair SV, Tamura H, Jayakumar R (2011) Efficient water soluble O-carboxymethyl chitosan nanocarrier for the delivery of curcumin to cancer cells. Carbohyd Polym 83(2):452–461.

Anton N, Vandamme TF (2009) The universality of low-energy nanoemulsification. Int J Pharmaceut 377(1):142–147.

Anton N, Benoit JP, Saulnier P (2008) Design and production of nanoparticles formulated from nano-emulsion templates: A review. J Control Release 128(3):185–199.

Anuchapreeda S, Fukumori Y, Okonogi S, Ichikawa H (2012) Preparation of Lipid Nanoemulsions Incorporating Curcumin for Cancer Therapy. J Nanotech, doi:10.1155/2012/270383.

C. Anandharamakrishnan, *Techniques for Nanoencapsulation of Food Ingredients*,
SpringerBriefs in Food, Health, and Nutrition, DOI 10.1007/978-1-4614-9387-7,
© C. Anandharamakrishnan 2014

Arya N, Chakraborty S, Dube N, Katti DS (2008) Electrospraying: A Facile Technique for Synthesis of Chitosan-Based Micro/Nanospheres for Drug Delivery Applications. Wiley Periodicals, Inc.

Attama AA, Momoh MA, Builders PF (2012) Lipid Nanoparticulate Drug Delivery Systems: A Revolution in Dosage Form Design and Development. InTech, Croatia, p 107–140.

Augustin MA, Hemar Y (2009) Nano- and micro-structured assemblies for encapsulation of food ingredients. Chem Soc Rev 38(4):902–912.

Augustin MA, Sanguansri P (2009) Nanostructured materials in the food industry. Adv Food Nutr Res 58:183–213.

Baker JR, Hamouda T, Shih A, Andrzej M (2003) Non-toxic antimicrobial compositions and methods of use. US Patent 6559189.

Bangham AD, Standish MM, Watkins JC (1965) Diffusion of Univalent Ions across the Lamellae of Swollen Phospholipids. J Mol Biol 13(1):238–252.

Bhandari BR, Datta N, Howes T (1997) Problems associated with spray drying of sugar-rich foods. Dry Technol 15(2):671–684.

Bhavsar RS, (2011) Overview of Homogenizer Processes. Pharma Times 43(10):37–41.

Bawa R, Bawa TSR, Maebius SB, Flynn T, Wei C (2005) Protecting new ideas and inventions in nanomedicine with patents. Nanomed-Nanotechnol 1(2):150–158.

Bejrapha P, Min SG, Surassmo S, Choi MJ (2010) Physicothermal properties of freeze-dried fish oil nanocapsules frozen under different conditions. Dry Technol 28(4):481–489.

Bejrapha P, Surassmo S, Choi M, Nakagawa K, Min S (2011) Studies on the role of gelatin as a cryo- and lyo-protectant in the stability of capsicum oleoresin nanocapsules in gelatin matrix. J Food Eng 105(2):320–331.

Berger N, Sachse A, Bender J, Schubert R, Brandl M (2001) Filter extrusion of liposomes using different devices: comparison of liposome size, encapsulation efficiency, and process characteristics. Int J Pharmaceut 223(1):55–68.

Bouwmeester H, Dekkers S, Noordam MY, Hagens WI, Bulder AS, de Heer C, Ten Voorde SE, Wijnhoven SW, Marvin HJ, Sips AJ (2009) Review of health safety aspects of nanotechnologies in food production. Regul Toxicol Pharm 53(1):52–62.

Brahatheeswaran D, Mathew A, Aswathy RG, Nagaoka Y, Venugopal K, Yoshida Y, Maekawa, T, Sakthikumar D (2012) Hybrid fluorescent curcumin loaded zein electrospun nanofibrous scaffold for biomedical applications. Biomedical Materials, doi: 10.1088/1748-6041/7/4/045001

Buranasuksombat U, Kwon YJ, Turner M, Bhandari B (2011) Influence of Emulsion Droplet Size on Antimicrobial Properties. Food Sci Biotechnol 20(3):793–800.

Cadena PG, Pereira MA, Cordeiro RB, Cavalcanti IM, Neto BB, Pimentel MCCB, Filho JLL, Silva VL, Santos-Magalhaes NS (2013) Nanoencapsulation of quercetin and resveratrol into elastic liposomes. Biochem Biophy Acta 1828(2): 309–316.

Canselier JR, Delmas H, Wilhelm AM, Abismail B (2002) Ultrasound emulsification – An overview. J Disper Sci Technol 23(1–3):333–349.

Casabona R, Escajedo CM, Epifanjo S, Emaldi L, Cirion A (2010) Safe and socially robust development of nanofood through ISO standards? Conference paper (EurSafe 2010, Bilbao, Spain, 16–81 September 2010) – Global food security: ethical and legal challenges 2010, p 521–526.

Chau CF, Wu SH, Yen GC (2007) The development of regulations for food nanotechnology. Trends Food Sci Tech 18(5):269–280.

Chaudhry Q, Castle L (2011) Food applications of nanotechnologies: An overview of opportunities and challenges for developing countries. Trends Food Sci Tech 10:1–9.

Cheftel CJ (2011) Emerging risks related to food technology. Advances in Food Protection. NATO Science for peace.

Chen MJ (2007) Development and Parametric Studies of Carbon Nanotube Dispersion Using Electrospraying. MS Thesis.

Chen LY, Remondetto GE, Subirade M (2006a) Food protein based materials as nutraceutical delivery systems. Trends Food Sci Tech 17(5):272–283.

Chen H, Weiss J, Shahidi F (2006b) Nanotechnology in nutraceuticals and functional foods. Food Technol 60(3):30–36.

Cheong JN, Tan CP, Yaakob B, Che M, Misran M (2008) α-Tocopherol nanodispersions: Preparation, characterization and stability evaluation. Int J Food Eng 89(2):204–209.

Chi-Fai C, Shiuan-Huei W, Gow-Chin Y (2007) The development of regulations for food nanotechnology. Trends Food Sci Tech 18(5):269–280.

Choi MJ, Ruktanonchai U, Min SG, Chun JY, Soottitantawat A (2010) Physical characteristics of fish oil encapsulated by ß-cyclodextrin using an aggregation method or polycaprolactone using an emulsion-diffusion method. Food Chem 119(4):1694–1703.

Choi MJ, Briancon S, Andrieu J, Min SG, Fessi H (2004) Effect of freeze-drying process conditions on the stability of nanoparticles. Dry Technol 22(1–2):335–346.

Chong GH, Yunus R, Abdullah N, Choong TSY, Spotar S (2009) Coating and encapsulation of nanoparticles using supercritical antisolvent. Am J Appl Sci 6(7):1352–1358.

Colas JC, Shi W, Rao VSN, Omri A, Mozafari MR, Singh H (2007) Microscopical investigations of nisin-loaded nanoliposomesprepared by Mozafari method and their bacterial targeting. Micron 38(8):841–847.

Couvreur P, Dubernet C, Puisieux F (1995) Controlled drug delivery with nanoparticles: current possibilities and future trends. Eur J Pharm Biopharm, 41(1):2–13.

Dandekar PP, Jain R, Patil S, Dhumal R, Tiwari D, Sharma S, Vanage G, Patravale V (2010) Curcumin-loaded hydrogel nanoparticles: application in anti-malarial therapy and toxicological evaluation. J Pharm Sci 99(12):4992–5010.

Dandekar P, Jain R, Kumar C, Subramanian S, Samuel G, Venkatesh M, Patravale V (2009) Curcumin loaded pH-sensitive nanoparticles for the treatment of colon cancer. J Biomed Nanotechnol 5(5):445–55.

De Kruif CG, Weinbreck F, DeVries R (2004) Complex coacervation of proteins and anionic polysaccharides. Curr Opin Colloid Interface Sci 9(5):340–349.

De Paz E, Martin A, Estrella A, Rodriguez-Rojo S, Matias AA, Duarte CMM, Cocero MJ (2012) Formulation of β-carotene by precipitation from pressurized ethyl acetate-on water emulsions for application as natural colorant. Food Hydrocolloid 26(1):17–27.

Devarajan V, Ravichandran V (2011) Nanoemulsions: As modified drug delivery tool. Int J Compr Pharm 2(4):1–6.

Domb AJ (1993) Lipospheres for controlled delivery of substances, United States Patent no: 5188837, 23 Feb, 1993.

Dong B, Smith ME, Wnek GE (2009) Encapsulation of Multiple Biological Compounds within a Single Electrospun Fiber. Small 5(13):1508–1512.

Donsi F, Annunziata M, Vincensi M, Ferrari G, (2011a) Design of nanoemulsion-based delivery systems of natural antimicrobials: Effect of the emulsifier. J Biotechnol 159(4):342–50

Donsi F, Annunziata M, Sessa M, Ferrari G (2011b) Nanoencapsulation of essential oils to enhance their antimicrobial activity in foods. LWT – Food Sci Technol 44(9):1908–1914.

Dowling AP (2004) Development of nanotechnologies. Mater Today 7(12):30–35.

Dube A, Ng K, Nicolazzo JA, Larson I (2010) Effective use of reducing agents and nanoparticle encapsulation in stabilizing catechins in alkaline solution. Food Chem 122(3):662–667.

Duncan VT (2011) Applications of nanotechnology in food packaging and food safety: Barrier materials, antimicrobials and sensors. J Colloid Interf Sci 363(1):1–24.

Dunn J (2004) A Mini Revolution. Food Manufacture, London, UK. http://www.foodmanufacture.co.uk/news/fullstory.php/aid/472/A%20mini%20revolution.htm

EFSA (2009) European Food Safety Authority. Scientific opinion on 'The potential risks arising from nanoscience and nanotechnologies on food and feed safety. Scientific opinion of the Scientific Committee. The EFSA J 958:1–39.

ETC Group. Action Group on Erosion, Technology and Concentration (2005). The potential impacts of nano-scale technologies on commodity markets: the implications for commodity dependent developing countries. Available from http://www.etcgroup.org/upload/publication/45/01/southcentre.commodities.pdf.

Ezhilarasi PN, Karthik P, Chhanwal N, Anandharamakrishnan C (2013) Nanoencapsulation Techniques for Food Bioactive Components: A Review. Food Bioprocess Tech 6(3):628–647.

Fang Z, Bhandari B (2010) Encapsulation of Polyphenols- a review. Trends Food Sci Tech 21(10):510–523.

Fathi M, Mozafari MR, Mohebbi M (2012) Nanoencapsulation of food ingredients using lipid based delivery system. Trends Food Sci Tech 23(1):13–27.

Fellows PJ (1998) Food processing technology-Principles and practice. Woodhead Publishing Limited, Cambridge.

Fernandez P, Andre V, Rieger J, Kuhnle A (2004) Nanoemulsion formation by emulsion phase inversion. Colloid Surf A 251(1):53–58.

Fernandez A, Torres-Giner S, Lagaron JM (2009) Novel route to stabilization of bioactive antioxidants by encapsulation in electrospun fibers of zein prolamine. Food Hydrocolloids 23(5): 1427–1432.

Ferreira I, Rocha S, Coelho M (2007) Encapsulation of Antioxidants by Spray-Drying. Chem Eng Trans 11(9):713–717.

Galindo-Rodriguez S, Allemann E, Fessi H, Doelker E (2004) Physicochemical parameters associated with nanoparticle formation in the salting-out, emulsification-diffusion and nanoprecipitation methods. Pharm Res 21(8):1428–39.

Gan Q, Wang T (2007) Chitosan nanoparticle as protein delivery carrier-Systematic examination of fabrication conditions for efficient loading and release. Colloid Surf B 59(1):24–34.

Gruere G, Narrod C, Abbott L (2011). Agricultural, Food, and Water Nanotechnologies for the Poor, Available at: http://www.ifpri.org/sites/default/files/publications/ifpridp 01064.pdf. Accessed 14 April 2013.

Gasco MR (1993) Method for producing solid lipid microspeheres having a narrow size distribution. US Patent 5,250,236, 9 Oct 1993.

Gasco MR (1997) Solid lipid nanospheres from warm micro-emulsions. Pharm Technol Eur 9:52–58.

Ghai D, Sinha VR (2012) Nanoemulsions as self-emulsified drug delivery carriers for enhanced permeability of the poorly water-soluble selective β 1-adrenoreceptor blocker Talinolol. Nanomed-Nanotechnol 8(5):618–26.

Gomez-Estaca J, Balaguer MP, Gavara R, Hernandez- Munoz P, (2012) Formation of zein nanoparticles by electrohydrodynamic atomization: Effect of the main processing variables and suitability for encapsulating the food coloring and active ingredient curcumin. Food Hydrocoll 28(1): 82–91.

Gou M, Men K, Shi H, Xiang M, Zhang J, Song J, Long J, Wan Y, Luo F, Zhao X, Qian Z, (2011) Curcumin-loaded biodegradable polymeric micelles for colon cancer therapy in vitro and in vivo. Nanoscale 3(4):1558–1567.

Gouin S (2004) Microencapsulation: industrial appraisal of existing technologies and trends. Trends Food Sci Tech 15(7):330–347.

Goyal P, Goyal K, Kumar SGY, Singh A, Ktare OP, Mishra DN (2005) Liposomal drug delivery systems- clinical applications. Acta pharmaceut 55(1):1–25.

Graveland-Bikker JF, De Kruif CG (2006) Unique milk protein based nanotubes: Food and nano-technology meet. Trends Food Sci Tech 17(5):196–203.

Grimm RL, Beauchamp JL (2005) Survey of Previous Research in Charged Particle Dynamics, Droplets in Electric fields and Electrospray Ionization. J Phys Chem B 109:8244.

Guterres SS, BeckII RCR, Pohlmann AR (2009) Spray drying technique to prepare innovative nanoparticlulated formulations for drug administration: a brief overview. Braz J Phy 39(1): 205–209

Gutierrez JM, Gonzalez C, Maestro A, Sole I, Pey CM, Nolla J (2008) Nano-emulsions: New applications and optimization of their preparation. Curr Opin Colloid Interface Sci 13(4): 245–251.

Hadaruga NG, Hadaruga DI, Paunescu V, Tatu C, Ordodi VL, Bandur G, Lupea AX (2006) Thermal stability of the linoleic acid/α- and β-cyclodextrin complexes. Food Chem 99: 500–508.

Han F, Li S, Yin R, Liu H, Xu L (2008) Effect of surfactants on the formation and characterization of a new type of colloidal drug delivery system: nanostructured lipid carriers. Colloids Surf A 315(1):210–216.

Hatanaka J, Kimura Y, Lai-Fu Z, Onoue S , Yamada S (2008) Physicochemical and pharmacokinetic characterization of water-soluble Coenzyme Q10 formulations. Int J Pharm 363(1–2): 112–117.

Hentschel A, Gramdorf S, Muller RH, Kurz T (2008) β-Carotene-Loaded Nanostructured Lipid Carriers. J food sci 73(2):N1-N6.

Herrera JE, Sakulchaicharoen N (2009) Microscopic and Spectroscopic Characterization of Nanoparticles, Informa Healthcare, New York, USA.

Jin H, Xia F, Jiang C, Zhao Y, He L (2009) Nanoencapsulation of lutein with hydroxypropylmethyl cellulose phthalate by supercritical antisolvent. Chinese J Chem Eng 17:672–677.

Heyang J, Fei X, Cuilan J, Yaping Z, Lin H (2009) Nanoencapsulation of lutein with hydroxypropylmethyl cellulose phthalate by supercritical antisolvent. Chinese Journal of Chemical Engineering 17(4):672–677.

Hindmarsh JP, Russell AB, Chen XD (2003) Experimental and numerical analysis of the temperature transition of a suspended freezing water droplet. Int J Heat Mass Tran 46(7):1199–1213.

Huang Q, Yu H, Ru Q (2010) Bioavailability and delivery of nutraceuticals using nanotechnology. J Food Sci 75:R50-R57.

Hughes GA (2005) Nanostructure-mediated drug delivery. Nanomed-Nanotechnol 1:22–30.

International Union of Food Science and Technology (2007). Nanotechnology and Food. Scientific Information bulletin.

Jafari SM, Assadpoor E, Bhandari B, He Y (2008) Nano-particle encapsulation of fish oil by spray drying. Food Res Int 41:172–183.

Jafari SM, He Y, Bhandari B (2007a) Production of sub-micron emulsions by ultrasound and microfluidization techniques. J Food Eng, 82:478–488.

Jafari SM, He Y, Bhandari B (2007b) Encapsulation of nanopartricles of D-limonene by spray drying: role of emulsifiers and emulsifying agent. Dry technol 25:1079–1089.

Jafari SM, He Y, Bhandari B (2006) Nano-emulsion production by sonication and Microfluidization-a comparison. Int J Food Prop 9:475–485.

Jaworek A (2007) Micro-and nanoparticle production by electrospraying. Powder Technol 176(1): 18–35.

Jenning V, Schafer-Korting M, Gohla S (2000) Vitamin A-loaded solid lipid nanoparticles for topical use: drug release properties. J Control Release 66:115–126.

Jincheng W, Xiaoyu Z, Siahao C (2010) Preparation and properties of nanoencapsulated capsaicin by complex coacervation method. Chemical Engineering Communication 197:919–933.

Joe, MM, Bradeeba K, Parthasarathi R, Sivakumaar PK, Chauhan PS, Tipayno S, Benson A, Sa T (2012a) Development of surfactin based nanoemulsion formulation from selected cooking oils: Evaluation for antimicrobial activity against selected food associated microorganisms. J Taiwan Inst Chemical Eng 43(2):172–180.

Joe MM, Chauhan PS, Bradeeba K, Shagol C, Sivakumaar PK, Sa T (2012b) Influence of sunflower oil based nanoemulsion (AUSN-4) on the shelf life and quality of Indo-Pacific king mackerel (*Scomberomorus guttatus*) steaks stored at 20° C. Food Control 23(2):564–570.

Kabalnov A, Wennerstrom H (1996) Macroemulsion stability: The oriented wedge theory revisited. Langmuir 12:276–92.

Karthikeyan R, Amaechi BT, Rawls HR, Lee VA (2012) Antimicrobial activity of nanoemulsion on cariogenic Streptococcus mutans. Arch Oral Biology 56(5):437–445.

Kayaci F, Uyar T (2012) Encapsulation of vanillin/cyclodextrin inclusion complex in electrospun polyvinyl alcohol (PVA) nanowebs: prolonged shelf-life and high temperature stability of vanillin. Food Chem 133(3): 641–649.

Kawashima Y (2001) Nanoparticulate system for improved drug delivery. Adv Drug Deliver Rev 47:1–2.

Kentish S, Wooster T, Ashokkumar M, Balachandran S, Mawson R, Simons L (2008) The use of ultrasonics for nanoemulsion preparation. Innov Food Sci Emerg Tech 9:170–175.

Khalil NM, do Nascimento TC, Casa DM, Dalmolin LF, de Mattos AC, Hoss I, Romano MA, Mainardes RM (2013) Pharmacokinetics of curcumin-loaded PLGA and PLGA–PEG blend nanoparticles after oral administration in rats. Colloid Surface B 101:353–360.

Khayata N, Abdelwahed W, Chehna MF, Charcosset C, Fessi H (2012) Preparation of vitamin E loaded nanocapsules by the nanoprecipitation method: From laboratory scale to large scale using a membrane contactor. Int J Pharmaceut 423:419–427.

Kikic I, Lora M, Bertucco A (1997) A thermodynamic analysis of three-phase equilibria in binary and ternary systems for applications in Rapid Expansion of a Supercritical Solution (RESS), Particles from Gas- Saturated Solutions (PGSS), and Supercritical Antisolvent (SAS). Ind Eng Chem Res 36:5507–5515.

King AH (1995) Encapsulation of Food Ingredients: A Review of Available Technology, Focusing on Hydrocolloids. In: Risch SJ, Reineccius GA (eds) Encapsulation and Controlled Release of Food Ingredients, ACS Symposium Series 590, American Chemical Society, Washington D.C. 26–39.

King CJ (1970) Freeze drying of foodstuffs. Critical Rev Food Technol 1–379.

Konan YN, Gurny R, Allemann E (2002) Preparation and characterization of sterile and freeze-dried sub-200 nm nanoparticles. Int J Pharmaceut 233(1):239–252.

Kuan CY, Yee-Fung W, Yuen KH, Liong MT (2012) Nanotech: Propensity in Foods and Bioactives. Crit Rev Food Sci 52:55–71.

Kuang SS, Oliveira JC, Crean AM (2010) Microencapsulation as a Tool for Incorporating Bioactive Ingredients into Food. Crit Rev Food Sci, 50:951–968.

Kumar MNVR, Kumar N (2001) Polymeric controlled drug-delivery systems: perspective issues and opportunities. Drug Dev Ind Pharm 27:1–30.

Kumari A, Yadav SK, Pakade YB, Singh B, Yadav SC (2010) Development of biodegradable nanoparticles for delivery of quercetin. Colloid Surfaces B 80(2):184–192.

Kumari A, Yadav SK, Pakade YB, Kumar V, Singh B, Chaudhary A, Yadav SC (2011) Nanoencapsulation and characterization of Albiziachinensis isolated antioxidant quercitrin on PLA nanoparticles. Colloid Surfaces B 82(1):224–232.

Kunieda H, Friberg SE (1981) Characterization of surfactants for enhanced oil recovery. Bull Chem Soc Jpn 54:1010.

Kuriakose R, Anandharamakrishnan C (2010) Computational fluid dynamics (CFD) applications in spray drying of food products. Trends Food Sci Tech 21(8):383–398.

Kuo F, Subramanian B, Kotyla T, Wilson TA, Yoganathan S, Nicolos RJ (2008) Nanoemulsions of an anti-oxidant synergy formulation containing gamma tocopherol have enhanced bioavailability and anti-inflammatory properties. Int J Pharm 363(1–2):206–213.

Kwon SS, Nam YS, Lee JS, Ku BS, Han SH, Lee JY, Chang IS (2002) Preparation and characterization of coenzyme Q10-loaded PMMA nanoparticles by a new emulsification process based on microfluidization. Colloid Surf A 210(1):95–104.

Lakkis JM (2007) Encapsulation and Controlled Release Technologies in Food Systems. Blackwell Publishing, Lowa.

Lawes G (1987) Scanning electron microscopy and X-ray microanalysis: Analytical chemistry by open learning, John Wiley & sons.

Leong TSH, Wooster TJ, Kentish SE, Ashokkumar M (2009) Minimising oil droplet size using ultrasonic emulsification. Ultrason Sonochem 16(6):721–727.

Leong WF, Lai OM, Long K, Yaakob B, Mana C, Misran M, Tan CP (2011). Preparation and characterisation of water-soluble phytosterol nanodispersions. Food Chem 129(1):77–83.

Levi G, Karel M (1995) Volumetric shrinkage (collapse) in freeze-dried carbohydrates above their glass transition temperature. Food Res Int 28(2): 145–151.

Li R, Qiao X, Li Q, He R, Ye M, Xiang C, Lina X, Guo D (2011) Metabolic and pharmacokinetic studies of curcumin, demethoxycurcumin and bisdemethoxycurcumin in mice tumor after intragastric administration of nanoparticle formulations by liquid chromatography coupled with tandem mass spectrometry. J Chromatogr B 879(26):2751–2758.

Li MK, Fogler HS (1978a) Acoustic emulsification. Part 1. The instability of the oil–water interface to form the initial droplets. J Fluid Mech 88(3):499–511.

Li MK, Fogler HS (1978b) Acoustic emulsification. Part 2. Break-up of the larger primary oil droplets in a water medium. J Fluid Mech 88(3):513–528.

Li W, Szoka F (2007) Lipid-based nanoparticles for nucleic acid delivery. Pharm Res 24(3):1–12.

Liang R, Huang Q, Ma J, Shoemaker CF, Zhong F (2013) Effect of relative humidity on the store stability of spray-dried beta-carotene nanoemulsions. Food Hydrocolloids 33(2):225–233.

Lippacher A, Muller RH, Mader K (2000) Investigation on the viscoelastic properties of lipid based colloidal drug carriers. Int J Pharmaceut 196:227–230.

Lippacher A, Muller RH, Mader K (2002) Semisolid SLN Dispersions for Topical Application: Influence of Formulation and Production Parameters on Viscoelastic Properties. Eur. J. Pharm. Biopharm 53(2):155–160.

Lira MCB, Ferraz MS, da Silva DGVC, Cortes ME, Teixeira KI, Caetano NP, Sinisterra RD, Santos-Magalhaes NS (2009) Inclusion complex of usnic acid with β-cyclodextrin: Characterization and nanoencapsulation into liposomes. J Incl Phenom Macro 64(3–4): 215–224.

Liu CH, Wu CT (2010) Optimization of nanostructured lipid carriers for lutein delivery. Colloid Surf A 353(2):149–156.

Liu N, Park H-J (2010) Factors effect on the loading efficiency of Vitamin C loaded chitosan-coated nanoliposomes Coll Surf B: Biointerfaces 76(1):16–19.

Liu J, Xu L, Liu C, Zhang D, Wang S, Deng Z, Lou W, Xu H, Bai Q, Ma J (2012a) Preparation and characterization of cationic curcumin nanoparticles for improvement of cellular uptake. Carbohyd Polymers 90(1):16–22.

Liu, G. Y., Wang, J. M., & Xia, Q. (2012b). Application of nanostructured lipid carrier in food for the improved bioavailability. Eur Food Res Technol 234:391–398.

Lobo L, Svereika A (2003) Coalescence during emulsification 2. Role of small molecule surfactants. J Colloid Interf Sci 261(2):498–507.

Lopez A, Gavara R, Lagaron J (2006) Bioactive packaging: turning foods into healthier foods through biomaterials. Trends Food Sci Tech 17(10):567–575.

Lopez-Rubio A, Sanchez E, Sanz Y, Lagaron JM (2009) Encapsulation of Living Bifidobacteria in Ultrathin PVOH Electrospun Fibers. Biomacromolecules 10(10):2823–2829.

Lopez-Rubio A, Lagaron JM (2012) Whey protein capsules obtained through electrospraying for the encapsulation of bioactives. Innov Food Sci Emerg 13:200–206.

Luo Y, Zhang B, Whent M, Yu L, Wang Q (2011) Preparation and characterization of zein/chitosan complex for encapsulation of α-tocopherol, and its in vitro controlled release study. Colloid Surf B 85:145–152.

Luykx DMAM, Peters RJB, Ruth SMV, Bouwmeester H (2008) A Review of Analytical Methods for the Identification and Characterization of Nano Delivery Systems in Food. J Agr Food Chem 56(18):8231–8247.

Machado AH, Lundberg D, Ribeiro AJ, Veiga FJ, Lindman B, Miguel MG, Olsson U (2012) Preparation of calcium alginate nanoparticles using water-in-oil (w/o) nanoemulsions. Langmuir 28(9):4131–4141.

Madene A, Jacquot M, Scher J, Desobry S (2006) Flavour encapsulation and controlled release – A review. Int J Food Science Technol 41(1):1–21.

Maherani B, Arab-Tehrany E, Kheirolomoom A, Cleymand F, Linder, M (2012) Influence of lipid composition on physicochemical properties of nanoliposomes encapsulating natural dipeptide antioxidant L-carnosine. Food Chem 134:632–640.

Manufuture (2006) Vision 2020 and Strategic Research Agenda of the European Agricultural Machinery Industry and Research Community for the 7th Framework Programme for Research of the European Community, Brussels, Belgium. http://www.manufuture.org/documents/ AET%20Vision%20and%20SRA1.pdf.

Mason TJ (1999) Sonochemistry. Oxford University Press, New York, USA.

Masters K (1991) Spray drying. Longman Scientific & Technical and John Wiley & Sons Inc., Essex, UK.

Mavrocordatos D, Pronk W, Boller M (2004) Analysis of environmental particles by atomic force microscopy, scanning and transmission electron microscopy. Water Sci Technol 50(12): 9–18.

McClements DJ, Decker EA, Park Y, Weiss J (2009) Structural design principles for delivery of bioactive components in nutraceuticals and functional foods. Crit Rev Food Sci 49(6): 577–606.

McClements DJ (2010) Emulsion design to improve the delivery of functional lipophilic components. In Ann Rev of Food Sci Technol 1:241–269.

McClements DJ, Rao J (2011) Food-Grade Nanoemulsions: Formulation, Fabrication, Properties, Performance, Biological Fate, and Potential Toxicity. Crit Rev Food Sci 51(4):285–330.

Mehnert W, Mader K (2012). Solid lipid nanoparticles Production, characterization and applications. Adv Drug Deliver Rev 64:83–101.

Meleson K, Graves S Mason T (2004) Formation of concentrated nanoemulsions by extreme shear. Soft Materials 2: 109–123.

Mishra B, Patel BB, Tiwari S (2010) Colloidal nanocarriers: a review on formulation technology, types and applications toward targeted drug delivery. Nanomed-Nanotechnol 6(1):9–24.

Mozafari MR, Flanagan J, Matia-Merino L, Awati A, Omri A, Suntres ZE, Singh H (2006) Recent trends in the lipid-based nanoencapsulation of antioxidants and their role in foods. J Sci Food Agr 86(13):2038–2045.

Mozafari MR (2005) Liposomes: an overview of manufacturing techniques. Cell Mol Biol Let 10(4):711–719.

Mozafari MR (2006) Bioactive entrapment and targeting using nanocarrier technologies: an introduction. In: Mozafari MR (ed) Nanocarrier technologies, Springer, Netherlands, p 1–16.

Mozafari MR, Khosravi-Darani K (2007) An overview of liposome-derived nanocarrier technologies. In: Mozafari MR (ed) Nanomaterials and nanosystems for biomedical applications, Springer, Dordrecht, p 113–123.

Mozafari MR, Mortazavi SM (eds) (2005) Nanoliposomes: from Fundamentals to Recent Developments. Trafford Pub. Ltd., Oxford, UK.

Mukerjee A, Vishwanatha JK (2009) Formulation, characterization and evaluation of curcumin loaded PLGA nanosphere for cancer therapy. J Anticancer Res 29(10):3867–3875.

Mukul D, Ansari KM, Anurag T, Dwivedi PD (2001) Need for Safety of Nanoparticles Used in Food Industry. J Biomed Nanotechnol 7(1):13–14.

Muller RH, Mader K, Gohla S (2000) Solid lipid nanoparticles (SLN) for controlled drug delivery – a review of the state of the art. Eur J Pharm Biopharm 50(1):161–177.

Muller RH, Mehnert W, Lucks JS, Schwarz C, Muhlen AZ, Weyhers H, Freitas C, Ruhl D (1995) Solid lipid nanoparticles (SLN) – An alternative colloidal carrier system for controlled drug delivery. Eur J Pharm Biopharm 41(1):62–69.

Muller RH, Radtke M, Wissing SA (2002) Solid lipid nanoparticles (SLN) and nanostructured lipid carriers (NLC) in cosmetic and dermatological preparations. Adv Drug Deliver Rev 54:S131-S155.

Nam YS, Kim Jin-W, Park J Y, Shim J, Lee JS, Han SH (2012) Tocopheryl acetate nanoemulsions stabilized with lipid–polymer hybrid emulsifiers for effective skin delivery. Colloids Surface B 94:51–57.

Nakagawa K, Surassmo S, Min SG, Choi MJ (2011) Dispersibility of freeze-dried poly(epsilon-caprolactone) nanocapsules stabilized by gelatin and the effect of freezing. J Food Eng 102(2):177–188.

Nayak AP, Tiyaboonchai W, Patankar S, Madhusudhan B, Souto EB (2010) Curcuminoids-loaded lipid nanoparticles: Novel approach towards malaria treatment. Colloid Surf B 81(1): 263–273.

Neethirajan S, Jayas DS (2010) Nanotechnology for the food and bioprocessing industries. Food Bioprocess Tech 4(1):39–47.

New RRC (1990) Introduction. In Liposomes: A Practical Approach, Oxford University Press, New York, USA.

Nobbmann U, Connah M, Fish B, Varley P, Gee C, Mulot S, Chen J, Zhou L, Lu Y, Sheng F, Yi J Harding SE (2007) Dynamic light scattering as a relative tool for assessing the molecular integrity and stability of monoclonal antibodies. Biotechnol Genet Eng Rev 24(1):117–128.

Oberdorster G, Maynard A, Donaldson K, Castranova V, Fitzpatrick J, Ausman K, Carter J, Karn B, Kreyling W, Lai D, Olin S, Riviere NM, Warheit D, Yang H (2005) Principles for characterizing the potential health effects from exposure to nano materials: elements of a screening strategy. Part Fibre Toxicol 2(1):8.

Olson DW, White CH, Richter RL (2004) Effect of pressure and fat content on particle sizes in microfluidized milk. J Dairy Sci 87(10):3217–3223.

Orive G, Anitua E, Pedraz JL, Emerich DF (2009) Biomaterials for promoting brain protection, repair and regeneration, Nat Rev Neurosci 10(9):682–692.

Oetjen GW (1999) Freeze-drying. Wiley-VCH, New York.

Pillai DS, Prabhasankar P, Jena BS, Anandharamakrishnan C (2012) Microencapsulation of *Garcinia* cowa fruit extract and effect of its use on pasta process and quality. Int J Food Prop 15(3):590–604.

Pinnamaneni S, Das NG, Das SK (2003) Comparison of oil-in water emulsions manufactured by Microfluidization and homogenization. Pharmazie 58(8):554–558.

Porras M, Solans C, Gonzalez C, Gutierrez JM (2008) Properties of water-in-oil (W/O) nano-emulsions prepared by a low-energy emulsification method. Colloid Surf A 324(1):181–188.

Pouton CW, Porter HCJ (2006) Formulation of Lipid-Based Delivery Systems for Oral Administration: Materials, Methods and Strategies. Annual Meeting of the American-Association-of-Pharmaceutical-Scientists, San Antonio, TX.

Powell JJ, Faria N, Thomas-McKay E, Pele CL (2010) Origin and fate of dietary nanoparticles and microparticles in the gastrointestinal tract. J Autoimmun 34(3):J226-J233.

Qian C, Decker EA, Xiao H, McClements DJ (2012) Physical and chemical stability of β-carotene-enriched nanoemulsions: Influence of pH, ionic strength, temperature, and emulsifier type. Food Chem 132(3):1221–1229.

Quintanar-Guerrero D, Allemann E, Fessi H, Doelker E (1998) Preparation techniques and mechanism of formation of biodegradable nanoparticles from preformed polymers. Drug Dev Ind Pharm 24(12):1113–1128.

Quintanilla-Carvajal MX, Camacho-Diaz BH, Meraz-Torres L S, Chanona-Perez JJ, Alamilla-Beltran L, Jimenez-Aparicio A, Gutierrez-Lopez GF (2010) Nanoencapsulation: A new trend in food engineering processing. Food Eng Rev 2(1):39–50.

Ragelle H, Crauste-Manciet S, Seguin J, Brossard D, Scherman D, Arnaud P, Chabot GG (2012) Nanoemulsion formulation of fisetin improves bioavailability and antitumour activity in mice. Int J Pharm 427(2):452–459.

Rasti B, Jina S, Mozafari MR, Yazid AM (2012) Comparative study of the oxidative and physical stability of liposomal and nanoliposomal polyunsaturated fatty acids prepared with conventional and Mozafari methods. Food Chem 135(4):2761–2770.

Rein MJ, Renouf M, Cruz-Hernandez C, Actis-Goretta L, Thakkar SK, Pinto MS (2012) Bioavailability of bioactive food compounds: a challenging journey to bioefficacy. Br J Clin Pharmacol 75(3):588–602

Reis CP, Neufeld RJ, Ribeiro AJ, Veiga F (2006) Nanoencapsulation I. Methods for preparation of drug-loaded polymeric nanoparticles. Nanomed-Nanotechnol 2(1):8–21.

Rejinold NS, Muthunarayanan M, Divyarani VV, Sreerekha PR, Chennazhi KP, Nair SV, Tamura H, Jayakumar R (2011) Curcumin-loaded biocompatible thermoresponsive polymeric nanoparticles for cancer drug delivery. J Colloid Interf Sci 360(1):39–51.

Relkin P, Yung JM, Kalnin D, Ollivon M (2008) Structural Behaviour of Lipid Droplets in Protein-stabilized Nano-emulsions and Stability of α-Tocopherol. Food Biophys 3(2):163–168.

Ribeiro HS, Chua BS, Ichikawab S, Nakajima M (2008) Preparation of nanodispersions containing β-carotene by solvent displacement method. Food Hydrocolloid 22(1):12–17.

Roco MC (2002) Nanoscale science and engineering for agriculture and food systems. Washington: National Planning Workshop, USDA/CSREES.

Rodgers TL, Cooke M, Hall S, Pacek A, Kowalski A (2011) Rotor-stator mixers. Chem Eng Trans 24:1411–1416.

Salvia-Trujillo L, Rojas-Grau A, Soliva-Fortuny R, Martin-Belloso O (2013) Effect of processing parameters on physicochemical characteristics of microfluidized lemongrass essential oil-alginate nanoemulsions. Food Hydrocolloid 30(1):401–407.

Sanguansri P, Augustin MA (2006) Nanoscale materials development-a food industry perspective. Trends Food Sci Tech 17(10):547–556.

Sathishkumar M, Sneha M, Won SW, Cho CW, Kim S, Yun YS (2009) Cinnamon zeylanicum bark extract and powder mediated green synthesis of nano-crystalline silver particles and its bactericidal activity. Colloid Surf B 73(2):332–338.

Shakeel F, Ramadan W (2010) Transdermal delivery of anticancer drug caffeine from water-in-oil nanoemulsions. Colloids Surf B: Biointerfaces 75(1):356–362.

Saupe A, Rades T (2006) Solid lipid nanoparticles. In: Mozafari MR (ed) Nanocarrier technologies. Frontiers of nanotherapy, Springer, Dordrecht, p 41–50.

Schubert H, Engel R (2004) Product and formulation engineering of emulsions. Chem Eng Res Des 82(9):1137–1143.

Schubert H, Ax K, Behrend O (2003) Product engineering of dispersed systems. Trends Food Sci Tech 14(1):9–16.

Schubert MA, Muller-Goymann CC (2005) Characterisation of surface-modified solid lipid nanoparticles (SLN): Influence of lecithin and nonionic emulsifier. Eur J Pharm Biopharmaceutics, 61: 77–86.

Sekhon BS (2010) Food nanotechnology – an overview. Nanotechnol Sci Appl 3:1–15.

Shaikh J, Ankola DD, Beniwal V, Singh D, Ravi Kumar, MNV (2009). Nanoparticle encapsulation improves oral bioavailability of curcumin at least 9-fold when compared to curcumin administered with piperine as absorption enhancer. Eur J Pharm Sci 37(3):223–230.

Shefer, A. (2008). The application of nanotechnology in the food industry. http://www.foodtech-international.com/papers/application-nano.htm.

Shegokar R, Muller RH (2010) Nanocrystals: industrially feasible multifunctional formulation technology for poorly soluble actives. Int J pharmaceut 399(1):129–139.

Shinoda K, Kunieda H (1983) Phase properties of emulsions: PIT and HLB. In: Becher P (ed) Encyclopedia of emulsion technology, Marcel Dekker, New York p 337–67.

Sill TJ, von Recum HA (2008) Electrospinning: Applications in drug delivery and tissue engineering. Biomaterials 29(13):1989–2006.

Silva HD, Cerqueira MA, Vicente AA (2012) Nanoemulsions for Food Applications: Development and Characterization. Food Bioprocess Tech 5(3):854–867.

Silva HD, Cerqueira MA, Souza BWS, Ribeiro C, Avides MC, Quintas MAC, Coimbra JSR, Cunha MGC, Vicente AA (2011) Nanoemulsions of β-carotene using a high-energy emulsification-evaporation technique. J Food Eng 102(2):130–135.

Singh RP, Heldman DR (4thed) (2009) Introduction to Food Engineering. Academic press, New York.

Solans C, Sole I (2012) Nano-emulsions: Formation by low-energy methods. Curr Opin Colloid In 17(5):246–254.

Sole I, Maestro A, Pey C, Gonzalez C, Solans C, Gutierrez JM (2006) Nano-emulsions preparation by low energy methods in an ionic surfactant system. Colloid Surf A 288(1):138–143.

Sonneville-Aubrun O, Babayan D, Bordeaux D, Lindner P, Rata G, Cabane B (2009) Phase transition pathways for the production of 100 nm oil-in-water emulsions. Phys Chem Chem Phys 11(1):101–10.

Sowasod N, Charinpanitkul ST, Tanthapanichakoon W (2008) Nanoencapsulation of curcumin in biodegradable chitosan via multiple emulsion/solvent evaporation. Int J Pharmaceut 347: 93–101.

Sozer N, Kokini JL (2009) Nanotechnology and its applications in the food sector. Trends Biotechnol 27(2):82–89.

Stang M, Schuchmann H, Schubert H (2001) Emulsification in high-pressure homogenizers. Eng Life Sci 1(4):151–157.

Stone WL, Smith M (2004) Therapeutic uses of antioxidant liposomes. Mol Biotechnol 27(3): 217–230.

Sugumar S, Nirmala J, Anjali H, Mukherjee A, Chandrasekaran N (2012) Bio-based nanoemulsion formulation, characterization and antibacterial activity against food-borne pathogens. J Basic Microbiol 52:1–10.

Sun XZ, Williams GR, Hou XX, Zhu LM (2013) Electrospun curcumin-loaded fibers with potential biomedical applications. Carbohyd Polym 94(1):147–153.

Suntres ZE, Shek PN (1996) Alleviation of paraquat-induced lung injury by pretreatment with bifunctional liposomes containing α-tocopherol and glutathione. Biochem Pharmacol 52(10): 1515–1520.

Surassamo S, Bejrapha P, Min SG, Choi MJ (2010) Effect of surfactants on capsicum oleoresin loaded nanocapsules formulated through emulsion diffusion method. Food Res Int 43(1):8–17.

Suwannateep N, Banlunara W, Wanichwecharungruang SP, Chiablaem K, Lirdprapamongkol K, Svasti J (2011) Mucoadhesive curcumin nanospheres: Biological activity, adhesion to stomach mucosa and release of curcumin into the circulation. J Control Release 151(2):176–182.

Suwantong O, Opanasopit P, Ruktanonchai U, Supaphol P (2007) Electrospun cellulose acetate fiber mats containing curcumin and release characteristic of the herbal substance. Polymer, 48(26): 7546–7557.

Suwantong O, Ruktanonchai U, Supaphol P (2010) In vitro biological evaluation of electrospun cellulose acetate fiber mats containing asiaticoside or curcumin. Journal of Biomed Mat Res Part A, 94(4): 1216–1225.

Tachaprutinun A, Udomsup T, Luadthong C, Wanichwecharungruang S (2009) Preventing the thermal degradation of astaxanthin through nanoencapsulation. Int J Pharmaceutics 374(1): 119–124.

Tadros T, Izquierdo R, Esquena J, Solans C (2004) Formation and stability of nano-emulsions. Adv Colloid Interface sci 108–109:303–318.

Taisne L, Cabane B (1998) Emulsification and Ripening following a Temperature Quench. Langmuir 14(17):4744–4752.

Taylor TM, Weiss J, Davidson PM, Bruce BD (2005) Liposomal nanocapsules in food science and agriculture. Crit Rev Food Sci 45(7–8):587–605.

Teeranachaideekul V, Müller RH, Junyaprasert VB (2007) Encapsulation of ascorbylpalmitate in nanostructured lipid carriers (NLC)-Effects of formulation parameters on physicochemical stability. Int J Pharmaceut 340(1):198–206.

Teska K, Kristl J (2010)The evidence for solid lipid nanoparticles mediated cell uptake of resveratrol. Int J Pharmaceut 390(1):61–69.

Thakur RK, Villette C, Aubry JM, Delaplace G (2008) Dynamic emulsification and catastrophic phase inversion of lecithin-based emulsions. Colloid Surf A 315(1):285–293.

Tice TR, Gilley RM (1985) Preparation of injectable controlled-release microcapsules by solvent-evaporation process. J Control Release 2:343–352.

Tiede K, Boxall A, Tear SP, Lewis J, David H, Hassellov M (2008) Detection and characterization of engineered nanoparticles in food and the environment. Food Addit Contam A 25(7):795–821.

Tiyaboonchai W, Tungpradit W, Plianbangchang P (2007) Formulation and characterization of curcuminoids loaded solid lipid nanoparticles. Int J Pharmaceut 337(1):299–306.

Tolstoguzov V (2003) Some thermodynamic considerations in food formulation. Food Hydrocolloid 17(1):1–23.

Torres-Giner S, Martinez-Abad A, Ocio MJ, Lagaron JM (2010) Stabilization of a Nutraceutical Omega-3 Fatty Acid by Encapsulation in Ultrathin Electro sprayed Zein Prolamine. J Food Sci 75(6):N69-N79.

Trotta M, Debernardi F, Caputo O (2003) Preparation of solid lipid nanoparticles by a solvent emulsification-diffusion technique. Int J Pharmaceut 257(1):153–160.

Tsai YM, Jan WC, Chien CF, Lee WC, Lin LC, Tsai TH (2011) Optimisednano-formulation on the bioavailability of hydrophobic polyphenol, curcumin, in freely-moving rats. Food Chem127:918–925.

Turgeon SL, Schmitt C, Sanchez C (2007) Protein–polysaccharide complexes and coacervates. Curr Opin Colloid Interface Sci 12(4):166–178.

Turk M, Lietzow R (2004) Stabilized Nanoparticles of Phytosterol by Rapid Expansion from Supercritical Solution Into Aqueous Solution. AAPS Pharm Sci Tech 5(4):1–10.

Uner M (2006) Preparation, characterization and physico-chemical properties of solid lipid nanoparticles (SLN) and nanostructured lipid carriers (NLC): their benefits as colloidal drug carrier systems. Pharmazie 61(5):375–386.

Varma MVS, Kaushal AM, Garg A, Garg S (2004). Factors affecting mechanism and kinetics of drug release from matrix-based oral controlled drug delivery systems. Am J Drug Deliv 2(1):43–57.

Venishetty VK, Chede R, Komuravelli R, Adepu L, Sistla R, Diwan PV (2012) Design and evaluation of polymer coated carvedilol loaded solid lipid nanoparticles to improve the oral bioavailability: A novel strategy to avoid intraduodenal administration. Colloid Surf B 95:1–9.

Venuganti VV, Perumal OP (2009) Nanosystems for Dermal and Transdermal Drug Delivery In: Drug Delivery Nanoparticles Formulation and Characterization Pathak Y Thassu D Informa Healthcare USA 2009 (126–155).

Vishwanathan R, Wilson TA, Nicolosi RJ (2009) Bioavailability of a Nanoemulsion of Lutein is Greater than a Lutein Supplement. Nano Biomed Eng 1(1):38–49

Walsh S Balbus MJ, Denison R, Florini K (2008) Nanotechnology: getting it right the first time. J Clean Prod 16:1018–1020.

Walstra P (1993) Principles of emulsion formation. Chem Eng Sci 48(2):333–349.

Walstra P (2003) Physical Chemistry of Foods. Marcel Decker, New York.

Wang JC, Chen SH, Xu ZC (2008a) Synthesis and properties research on the nanocapsulated capsaicin by simple coacervation method. J Disper Sci Technol 29(5):687–695.

Wang X, Jiang Y, Wang YW, Huang MT, Hoa CT, Huang Q (2008b) Enhancing anti-inflammation activity of curcumin through O/W nanoemulsions. Food Chem 108(2):419–424.

Weiner BB, Tscharnuter WW, Fairhurst D (1993) Zeta Potential: A New Approach. Canadian Mineral Analysts Meeting.

Weiss J, Takhistov P, Mcclements DJ (2006) Functional materials in food nanotechnology. J Food Sci 71(9) R107-R116.

Williams DB (1996) Transmission Electron Microscopy, A textbook for Material Science, Plenum Press, New York and London.

Wooster T, Golding M, Sanguansri P (2008) Impact of oil type on nanoemulsion formation and Ostwald ripening stability. Langmuir 24(22):12758–12765.

Wu Y, Clark RL (2008) Electrohydrodynamic atomization: a versatile process for preparing materials for biomedical applications. J Biomat Sci-Polym E 19(5):573–601.

Xia S, Xu S, Zhang X (2006) Optimization in the Preparation of Coenzyme Q10 Nanoliposomes. J Agric Food Chem 54(17:6358–6366

Xie X, Tao Q, Zou Y, Zhang F, Guo M, Wang Y, Wang H, Zhou Q, Yu SJ (2011) PLGA nanoparticles improve the oral bioavailability of curcumin in rats: characterizations and mechanisms. J Agr Food Chem 59(17):9280–9289.

Xing F, Cheng G, Yi K, Ma L (2004) Nanoencapsulation of capsaicin by complex coacervation of gelatin acacia, and tannins. J Appl Polym Sci 96(6):2225–2229.

Yallapu MM, Gupta BK, Jaggi M, Chauhan SC (2010) Fabrication of curcumin encapsulated PLGA nanoparticles for improved therapeutic effects in metastatic cancer cells. J Colloid Interf Sci 351(1):19–29.

Yin L, Chu B, Kobayashi I, Nakajima M (2009) Performance of selected emulsifiers and their combinations in the preparation of beta-carotene nanodispersions. Food Hydrocolloids 23(6):1617–1622.

Yuan Y, Gao Y, Zhao J, Mao L (2008) Characterization and stability evaluation of [beta]-carotene nanoemulsions prepared by high pressure homogenization under various emulsifying conditions. Food Res Int 41(1):61–68.

Zambaux M, Bonneaux F, Gref R, Maincent P, Dellacherie E, Alonso M, Labrude P, Vigneron C (1998) Influence of experimental parameters on the characteristics of poly(lactic acid) nanoparticles prepared by double emulsion method. J Control Release 50(1):31–40.

Zhang YZ, Wang X, Feng Y, Li J, Lim CT, Ramakrishn S (2006) Coaxial electrospinning of (fluorescein isothiocyanate-conjugated bovine serum albumin)-encapsulated poly (ε-caprolactone) nanofibers for sustained release Biomacromolecules 7(4):1049–1057.

Zhao L, Xiong H, Peng H, Wang Q, Han D, Bai C, Liu Y, Shi S, Deng B (2011) PEG-coated lyophilized pro-liposomes: preparation, characterizations and in vitro release evaluation of vitamin E. Eur Food Res Technol 232(4):647–654.

Zhou H, Yue Y, Liu G, Li Y, Zhang J, Gong Q, Yan Z, Duan M (2010) Preparation and Characterization of a Lecithin Nanoemulsion as a Topical Delivery System. Nanoscale Res Lett 5(1):224–230.

Zhuang CY, Li N, Wang M, Zhang XN, Pan WS, Peng JJ, Pan YS, Tang X (2010) Preparation and characterization of vinpocetine loaded nanostructured lipid carriers (NLC) for improved oral bioavailability. Int J Pharmaceut 394(1):179–185.

Zimet P, Livney YD (2009) Beta-lactoglobulin and its nanocomplexes with pectin as vehicles for ω-3 polyunsaturated fatty acids. Food Hydrocolloid 23(4):1120–1126.

Zuidam NJ, Shimoni E (2010) Overview of microencapsulation use in food products or processes and methods to make them. In: Zuidam NJ, Nedovic VA (eds) Encapsulation technique for active food ingredients and food processing, Springer, NewYork, 3–29.

Index

A
AFM. *See* Atomic force microscopy (AFM)
Aggregation, 6, 57, 58, 60, 70
Anti-solvent precipitation, 31, 40
Astaxanthin, 30, 31, 33, 34, 37
Atomic force microscopy (AFM), 6, 68, 69
Auger electron spectroscopy, 69

B
Backscattered electrons (BSE), 68
Bangham thin-film hydration, 25
β-cyclodextrin (β-CD), 30, 38, 44, 48, 57, 59
Bifidobacteria, 44, 48
Bifunctional liposomes, 23
Bioactive compounds, 1–9, 17–19, 25, 27, 29–31, 40, 44, 48, 49, 59, 62–64
Bioactive entrapment, 17–28
Bioavailability, 2, 4, 5, 19, 20, 27, 28, 30–32, 34, 35, 54, 56, 59, 62, 64–65, 72
Biodegradable, 5, 29, 48, 56
Biopersistent, 72
Bottom-up approaches, 5, 6
BSE. *See* Backscattered electrons (BSE)

C
Capsaicin, 30, 39
Capsicum oleoresin, 58, 59
β-Carotene, 4, 7, 9–12, 20, 26, 30, 31, 33, 37, 44, 46, 49, 54, 59
Catechin, 53, 54, 59, 60
Charge-coupled device, 69
Chitosan, 5, 10, 25, 30–32, 34, 40, 44, 45, 56, 59, 60
Coacervation, 5, 30, 38–40

D
DHA. *See* Docosahexaenoic acid (DHA)
ᴅ-Limonene, 9, 11, 13, 59, 62
DLS. *See* Dynamic light scattering (DLS)
Docosahexaenoic acid (DHA), 4, 20, 25, 38, 45
Dose metrics, 74
Droplet diameter, 9, 12, 33, 37, 65
Dynamic light scattering (DLS), 68

Codex Alimentarius, 73
Coenzyme, 19, 20, 25, 27, 31, 36, 64
Core material, 5, 23, 44, 55
Cryoprotectant, 58, 60
Curcumin, 9, 11, 19, 22, 27, 29–32, 34, 35, 44, 46, 48, 49, 56, 59, 64, 65
Curcuminoids, 19, 22, 59

E
EFSA. *See* European Food Safety Authority (EFSA)
EIP. *See* Emulsion inversion point (EIP)
Electro hydrodynamic atomization, 43
Electrospinning, 41, 43–49
Electro-spraying, 41, 43–49
Electrostatic repulsion, 70
Emulsification, 5, 8, 9, 11, 13–16, 22, 29, 31, 34–37, 41, 54, 55, 60
Emulsification-solvent evaporation, 5, 6, 34–37
Emulsion inversion point (EIP), 16
Entrapment, 17–28
European Food Safety Authority (EFSA), 73, 74
Extrusion, 24

C. Anandharamakrishnan, *Techniques for Nanoencapsulation of Food Ingredients*,
SpringerBriefs in Food, Health, and Nutrition, DOI 10.1007/978-1-4614-9387-7,
© C. Anandharamakrishnan 2014